SANS DEMYSTIFIED

SANs Demystified

Denise Colon

McGraw-Hill
New York   Chicago   San Francisco   Lisbon
London   Madrid   Mexico City   Milan   New Delhi
San Juan   Seoul   Singapore   Sydney   Toronto

The McGraw·Hill Companies

Cataloging-in-Publication Data is on file with the Library of Congress.

Copyright © 2003 by The McGraw-Hill Companies, Inc. All rights reserved. Printed in the United States of America. Except as permitted under the United States Copyright Act of 1976, no part of this publication may be reproduced or distributed in any form or by any means, or stored in a data base or retrieval system, without the prior written permission of the publisher.

1 2 3 4 5 6 7 8 9 0 DOC/DOC 0 9 8 7 6 5 4 3 2

ISBN 0-07-139658-6

The sponsoring editor for this book was Judy Bass and the production supervisor was Sherri Souffrance. It was set in Century Schoolbook by MacAllister Publishing Services, LLC.

Printed and bound by RR Donnelley.

 This book is printed on recycled, acid-free paper containing a minimum of 50 percent recycled de-inked fiber.

McGraw-Hill books are available at special quantity discounts to use as premiums and sales promotions, or for use in corporate training programs. For more information, please write to the Director of Special Sales, Professional Publishing, McGraw-Hill, Two Penn Plaza, New York, NY 10121-2298. Or contact your local bookstore.

Information contained in this work has been obtained by The McGraw-Hill Companies, Inc. ("McGraw-Hill") from sources believed to be reliable. However, neither McGraw-Hill nor its authors guarantee the accuracy or completeness of any information published herein, and neither McGraw-Hill nor its authors shall be responsible for any errors, omissions, or damages arising out of use of this information. This work is published with the understanding that McGraw-Hill and its authors are supplying information but are not attempting to render engineering or other professional services. If such services are required, the assistance of an appropriate professional should be sought.

To my children, Gina, Lana and Karl and my grandchildren, Anissa and Nicholas. Thanks for all of your support.

CONTENTS

Chapter 1 Understanding Storage Area Networks 1

    What's in a SAN? 3
        SAN Interfaces 3
        SAN Interconnects 3
        SAN Fabrics 4
    Who Needs a SAN? 4
    What Should Every Good SAN Do? 4
    Designing a SAN 5
    The Preferred Technology That Powers a SAN 6
    Summary 6

Chapter 2 Understanding How a SAN Works 7

    The Basics 8
        The Fibre Channel SAN Fabric 8
    How Fibre Channel Works 9
    Protocol Basics to SAN Services 13
        FC-0 Layer: Physical Layer 13
        FC-1 Layer: Transmission Encode/Decode Layer 13
        FC-2 Layer: Framing and Signaling Layer 13
        FC-3 Layer: Advanced Features 13
        FC-4 Layer: Protocol Mapping 14
    Fibre Channel Topologies 15
        Point-to-Point 15
        Arbitrated Loop 15
        Fabric Switched 16
    Fibre Channel Architectures by Design 17
        8/2/96 Architecture 18
        16/4/192 Architecture 19
        16/2/224 Architecture 20
        12/2/160 Architecture 21
    Other Emerging Technologies 22
    Summary 24

Chapter 3	SAN Setup	25
	Combining Switch Access and Shared Media	26
	Managing a SAN	28
	Common SAN Terminology and Organizations	29
	Understanding the Difference Between NAS and SAN	31
	Summary	32
Chapter 4	Configuring SANs: Dos and Don'ts	33
	Configuring from Existing Legacy Systems	35
	SAN Configuration	38
	System Requirements and Analysis	38
	SAN Architecture	40
	Building a Testbed	40
	Legacy-to-SAN Environment	41
	Storage Performance Survey	44
	Storage Architecture and Design	45
	Return on Investments	47
	Interoperability Issues	51
	SAN Gateways	53
	Project Management	55
	Examples of Successful SAN Transitions	56
	Peer Model	57
	Dense Wavelength Division Multiplexing	58
	The Bandwidth Crisis	58
	Possible Solutions to the Bandwidth Crisis	60
	What Is DWDM?	61
	Is DWDM Flexible?	62
	Is DWDM Expandable?	63
	DWDM Costs	64
	Metropolitan Area Networks (MANs)	67
	Wide Area Newtorks (WANs)	69
	Asynchronous Transfer Mode (ATM)	73
	Summary	76
Chapter 5	Fibre Channel	79
	Fibre Channel Overview	80
	Concepts and Terminology	81

Contents

Fibre Channel Architecture	91
FC-0 Layer: Physical Layer	92
FC-1 Layer: Transmission Encode/Decode Layer	92
FC-2 Layer: Framing and Signaling Layer	93
FC-3 Layer: Advanced Features	94
FC-4 Layer: Protocol Mapping	95
Ports	95
Fibre Channel Service Classes	96
Class 1	97
Class 2	97
Class 3	98
Class 4	99
Class 5	99
Class 6	100
Intermix	100
Fibre Channel Topologies	100
Point to Point	100
Arbitrated Loop	101
Fabric	104
More Concepts and Terminology	104
Copper Cables	104
Disk Enclosures	104
Drivers	105
Extenders	105
Fibre Channel Disks	105
Fibre-Optic Cable Connector	105
Gigabit Interface Converters	105
Gigabit Link Modules	106
Host Bus Adapter (HBA)	106
Hubs	106
Link Analyzer	107
Multimode Cable	107
Routers: LAN Switch	107
SCSI Bridge	107
Static Switches	107
Switch WAN Extender	108
Addressing	108
Login	109

	Transmission Hierarchy	110
	8B/10B Transmission Character	110
	Transmission Word	111
	Frame	111
	Sequence	112
	Exchange	112
	Summary	113
Chapter 6	Case Studies	115
	SAN Implementation Saves Money and Boosts Network	
	Performance at New York Life	116
	In the Beginning . . .	116
	The Solution	117
	Saving Big Bucks	118
	A By-Product: Complexity	118
	Analyzing Storage Needs	119
	The Value of an Independent Consultant	119
	R&B Group Adds Redundancy and Speed	
	with Fibre Channel Adapter Cards	120
	Overcoming Major Drawbacks	121
	Improved System Access	121
	Forest Products Manufacturer Takes Control of Its Storage	122
	Cleanup Campaign	123
	Backup Time Savings	124
	Two Risk-Averse Users Tread Cautiously During	
	SAN Development	125
	Gary Fox: Developing a Nationwide SAN	125
	Michael Butler: Seeking Business Continuity	126
	Law Firm Backs Up Data to "EVaults"	127
	The Solution: EVault's Online Backup	128
	Global Provider Relies On Flexible Storage	129
	The Challenge: High-Availability Storage	130
	The Solution: Storage on Demand	131
	Important Criteria	131
	The Benefits: Economy, Flexibility	132
	Data Restoration Package One-Ups Lovebug and	
	Anna Kournikova	133
	Dealing with the Anna Kournikova Virus	135

Contents

Euroconex Turns to MTI Vivant for High-Availability Data Storage	136
The Solution: MTI Vivant	137
Disaster Recovery	138
Achieving High Availability	138
With $100 Trillion in the Balance, DTCC Plays It Close to the Vest	139
Ensuring Redundancy	139
Seeking Safety and Soundness	140
Boosting Backup Speed	141
Going for a High Level of Support	142
SAN-Bound A. B. Watley on Migration Path with Sun StorEdge T3 Arrays	143
Selecting the Right Sun Hardware Solutions	144
Planning for SAN	146
New SNIA Technology Center Opened for Interoperability Testing	147
Summary	149

Glossary 151

Appendix A Quick Reference Card 265

Index 269

CHAPTER 1

Understanding Storage Area Networks

What is a *storage area network* (SAN)? A SAN is a high-speed dedicated network that is not unlike a *local area network* (LAN). A SAN establishes direct connections between storage elements, clients, or servers. SANs are developed through the use of multiple storage devices (such as a *redundant array of independent disks* [RAID], *just a bunch of disks* [JBODs], or tape libraries) that are connected in an any-to-any relationship and accessed via one or more servers. In plain English, SAN systems do not require server connections; they are LAN-free backup systems. They do not need to be housed in the same box as servers, nor are they required to be from the same manufacturing companies as servers. Rather than putting data storage directly on the network, the SAN solution puts data storage network devices *between* storage subsystems and the servers. SANs can be built as switched- and/or shared-access networks. They offer exceptional improvements in scalability, fault recovery, and diagnostic analysis information.

Why is data storage such an important issue? Well, it is estimated that 3.2 million *exabytes* of information exist on the earth today, and this number exceed 43 million by the year 2005. And in case you're wondering, an *exa* is defined as 1 billion, so in decimal terms, one exabyte is equal to a billion gigabytes! With data, or information, constituting such a large measurement, it is no surprise that data storage has become an issue of major importance for modern businesses.

The benefits that SANs offer when addressing data storage issues are that they provide environments where

- All data storage assets are shared among multiple servers without being physically attached to their storage bases.
- Data movements and manipulation are being managed with greater ease and efficiency.
- There are significant reductions in the negative impacts on critical business applications.

In short, SANs have been developed to alleviate the limitations of single-server storage. They are, for all intents and purposes, extended storage buses that can be interconnected using similar interconnect technologies that are typically already used on LANs and *wide area networks* (WANs), namely, routers, hubs, switches, and gateways.

What's in a SAN?

What comprises a SAN? Though commonly spoken of in terms of hardware, SANs also include specialized software for managing, monitoring, and configuring data storage. In discussing SAN hardware, we refer to three major components:

- Interfaces-*Host bus adapters* (HBAs)
- Interconnects-Targets
- Fabrics-Switches

SAN Interfaces

Common SAN Interfaces include Fibre Channel, *Small Computer System Interface* (SCSI), SSA, *Enterprise Systems Connection* (ESCON), *High-Performance Parallel Interface* (HIPPI), and Bus-&-Tags. All these options allow the storage of data to exist externally to the server. They also can host shared storage configurations for clustering.

SAN Interconnects

Examples of SAN interconnects include multiplexors, hubs, routers, extenders, gateways, switches, and directors. Typically, *information technology* (IT) professionals are familiar with these terms if they have installed a LAN or WAN. The SAN interconnect ties a SAN interface into a network configuration across large distances. The interconnect also links the interface to SAN fabrics.

SAN Fabrics

Switched SCSI, Fibre Channel switched, and switched *Serial Storage Architecture* (SSA) are the most common SAN fabrics.

Who Needs a SAN?

One of the biggest problems most businesses face is how to store and manage their data. SANs enable applications that move data around the network to perform optimally by adding bandwidth for specific functions without placing a load on primary networks. SANs enable data warehousing and other higher-performance solutions. Increasingly, almost any small, midsized, or large businesses can justify the need for a SAN, particularly businesses that require quick access to information. Businesses that are involved in transaction-based applications, such as e-commerce, customer service, and/or financial applications, benefit greatly by using a SAN. Likewise, businesses in which transaction volumes are often unpredictable require SANs.

Additionally, businesses in which customers require 24/7 availability would benefit tremendously from a SAN solution. The need for SANs is based on workload, and with the ever-increasing use of *network technology* (NT) for enterprise applications, SANs ultimately will be found supporting many Windows NT applications. SANs frequently support UNIX and other mainframe operating systems as well. These operating systems offer the reliability required for 24/7 availability.

What Should Every Good SAN Do?

It is apparent that given the increased storage capacity requirements and the fact that there are increasingly more users attempting to gain access to stored files, businesses need systems that can scale without creating further management issues. They require uninterrupted application availability in addition to faster access of their stored

Understanding Storage Area Networks

data. A SAN disaster-recovery detail report (along with a business continuity plan) can help to justify investments in SANs. Downtime losses per hour vary by business and industry, but the hourly average downtime loss is approximately $84,000. A generic SAN can be implemented for approximately $150,000 contingent on costs per port.

Many people agree that SAN architectures offer the best solutions to ensure ongoing business operations and continuity. Therefore, what should we look for in a well-configured SAN? A good SAN should provide greater capacity and be able to store more than 500 GB of data. A good SAN also should provide data availability and faster access.

Designing a SAN

Not every SAN is alike. While designing your SAN, of course, you'll want it to address your specific needs. However, there are a few ground rules. A well-designed SAN will achieve the following:

Externalized storage to gain the benefits of improved performance. This will provide freedom from server outages, higher data accessibility, higher system availability, and easier management at a lower cost.

Centralized storage repositories such as backup, archives, references, documents, data warehouses, and/or other shared resources. This will lower administrative costs and improve data access.

Remote clustering solutions. This will establish real-time dynamic business continuity rather than just "throwing data over the wall."

More specifically, the major features that a well-designed SAN offers include

- High bandwidth—2 Gbps
- Disaster recovery plans
- Business continuity plan
- Manageability

- Easy integration
- Lower total cost of ownership

The Preferred Technology That Powers a SAN

Presently, Fibre Channel is used because it has characteristics that are suited to SCSI and storage. The ability to pool storage resources in order to increase the accessibility and manageability of data is the principal advantage of a Fibre Channel SAN. The following comparison puts this nicely into perspective: Imagine that you have a file that is represented by a stack of 500 sheets of paper. With Fibre Channel, I simply would hand you the stack for processing. Using *Transmission Control Protocol / Internet Protocol* (TCP/IP), for example, I would give you the stack one sheet at a time!

Summary

To summarize, a SAN is a high-speed dedicated network, not unlike a LAN, that establishes a direct connection between storage elements, clients, or servers. SANs rely on Fibre Channels for higher throughput, greater distancing, and enhanced connectivity options between servers and storage devices. They can be built as switched- or shared-access networks, and they offer improved scalability, fault recovery, and diagnostic analysis information while maintaining higher availabilities than current approaches. Although typically spoken of in terms of hardware, SANs include specialized software to manage, monitor, and configure system data storage. Additionally, in the design of a SAN, we know that important considerations include high bandwidth, a disaster recovery plan, a business continuity plan, easy manageability and integration, and lower total cost of ownership. In Chapter 2, "Understanding How a SAN Works," we will delve into just how a SAN works and the recent improvements that have advanced SAN technology.

CHAPTER 2

Understanding How a SAN Works

The Basics

The Fibre Channel SAN Fabric

Fibre Channel is the power behind the SAN. Fibre Channel allows for an active intelligent interconnection scheme, called a *fabric*, to connect devices. In the context of a *storage area network* (SAN), the fabric typically refers to the detailed makeup of the network, such as cards and attached devices. The most common SAN fabrics include Fibre Channel switched, switched *Small Computer System Interface* (SCSI), and switched *Serial Storage Architecture* (SSA). Generally, SAN interconnects are connected to Fibre Channel switches. Switches allow many advantages in building centralized, centrally managed, and consolidated storage repositories shared across a number of applications. The switches furnish the backbone for all the connected devices, with one or more of the switches acting as a Fibre Channel switching fabric. SAN switch fabrics allow attachments of thousands of nodes. Although SAN fabrics are often mixed together, it should be remembered that they are really distinct elements of the SAN.

Fibre Channel architecture offers several topologies for network design, primarily point-to-point, arbitrated loop, and switch. The ports in a point-to-point connection are called *N_Ports*, loop connections are called *NL_Ports*, and a Fibre Channel switch or network of switches is called a *fabric*. All are based on gigabit speeds, with effective 100 MBps throughput (200 MB full duplex). All allow for both copper and fiberoptic cable plant, with maximum distances appropriate to the media (30 m for copper, 500 m for short-wave laser over multimode fiber, 10 km for long-wave laser over single-mode fiber). Fibre Channel topologies are discussed in greater detail in Chapter 5, "Fibre Channel," including a comparison chart listing the pros and cons of each topology.

How Fibre Channel Works

Two serial cables, one carrying inbound data and the other carrying outbound data, make the basic connection to a Fibre Channel device. These cables can be fiberoptic or twin-axial copper. A Fibre Channel installation has a minimum of one link between two nodes. As mentioned earlier, the data flow between hardware entities called *N_ports,* generally part of the termination card, which contains hardware and software support for the Fibre Channel protocol. Each node must have at least one N_port and usually has two; either may serve as transmitter, receiver, or both. An N_port is assigned a unique address (the N_port identifier) and also contains a *link control facility* (LCF) that handles the logical and physical control of the link at the firmware level. Everything between the ports on Fibre Channel is called the *fabric.* The fabric is most often a switch or series of switches that takes the responsibility for routing. Ports on the fabric are called *F_ports.* All a Fibre Channel node has to do is manage a simple point-to-point connection between itself (N_port) and the fabric (F_port). NL_port and FL_port refer to an N_port and F_port, respectively, that can support arbitrated loop functions in addition to basic point-to-point functions.

For a real appreciation of advancements that have been made regarding storage technology (the movement of data from servers to storage devices), a quick analysis of upgrade history in this area is required. Table 2-1 shows typical storage options that customers have used in the past.

As with all data transport technologies of the past, the goal of achieving data transport from point *A* to point *B* with thorough speed, integrity, and efficiency is desired. When using SANs, all points *A* and *B* points are known as (*A*) the *initiator* (or host bus adapter) and (*B*) the *target* (or disk). (It is essential to note here that there are predefined rules that data sets must follow to arrive at their destinations in one piece.)

For the sake of simplicity, let's reference a scenario where people are being transported by automobile. In our scenario, a father is dri-

Table 2-1

The history of storage technology

Technology	MBps	Max. Devices	SCSI Cable Lengths
SCSI 21	5	8	25 m
SCSI 22	10	8	25 m
FW SCSI 22	20	16	25 m
Wide Ultra SCSI	40	16	25 m
Wide Ultra 2 SCSI	80	16	25 m
Ultra 3 SCSI	160	16	25 m
SSA	20–60	127	15 m
ESCON	4–16		
Fibre Channel	100	16 million	10,000 m
HIPPI	800	2 PTP	
FC/IP	1,000		

ving to his mother's house with his young children. His rules for travel are predefined to ensure that everyone in his automobile will arrive at the destination (grandmother's house) safely. These predefined rules communicate the requirements that the father needs to follow. Some of these requirements are the use of seat belts, the observation of speed limits, a valid license in the driver's possession, an agreement to obey road signs and speed limits, the implementation and use of turn signals, and the enactment of a multitude of other regulations involved in safe commuting.

In our example, the children signify the payload, or end result. Obeying the rules of the road will result in a successful passenger (or data) payload with a safe arrival. Neglect, or misuse, of one or any of the predefined rules will result in an unsafe journey in which the children may not arrive intact at grandma's home. An unsafe journey would create a failed data payload, and the result would be a failed data arrival. An unsafe journey result for either children or data is, of course, tragic. While children in this example are certainly of greater importance, certainly business data also carry a significant

Understanding How a SAN Works

weight of importance. Data must arrive at its destination. SANs provide this safe arrival.

The *Fibre Channel* is a protocol, or a set of rules. With Fibre Channels, the transportation medium may be copper or fiberoptic cabling. The protocols (signal rules) that use copper or fiberoptic cabling are broken down into Fibre Channel layers. Fibre Channels are used to logically identify the stack data must pass through to make communication complete. This layering perspective helps significantly when completing *Fibre Channel protocols* (FCPs), software troubleshooting, and/or SAN troubleshooting.

Additionally, when comparing FCPs to father's car ride, you may note that the car drives within a single lane of traffic when moving to and from the destination. Likewise, the Fibre Channel maintains this same concept. Its transportation is a serial transmission in single-line fashion. One line transmits, and a second line receives. The packets are lined up and transported in a serial approach.

Historically, Fibre Channel technology emerged from network attribute extractions that combined with channeled connection characteristics. A *channel* is a hardware-intensive high-speed data pipe that maintains a high degree of reliability. A channel, more accurately, SCSI III Fibre Channel, is designed to support speed and reliability of channeling protocols; SCSI and/or *High-Performance Parallel Interfaces* (HIPPIs) achieve this with network flexibility through the addition and removal of devices in an accelerated manner. Additionally, the Fibre Channel specification supports the *local area network* (LAN) and *wide area network* (WAN) protocols of *Internet Protocol* (IP), *Fiber Distributed Data Interface* (FDDI), and *Asynchronous Transfer Mode* (ATM). Primary devices used in SAN environments are initiating devices, target devices, switches, bridges, and hubs. They are classified as *interconnecting SAN devices* because they tie all the components together. The initiating device is a device that looks for and communicates with target devices. It is commonly referred to as the *host bus adapter* (HBA), and it resides in the server or client workstation. The significance of the HBA is that it actually is an active device that seeks out its targeted pair to communicate with so as to achieve a file transfer. The HBA has the ability to monitor its path to its targeted pair. If, for some reason, it loses contact with the target, the initiator will switch to an alternate target. In order for this event to occur, one level of redundancy is put in place whereby there are dual initiators, dual switches, and mirrored disks. Figure 2-1 depicts an entry-level HBA. Figure 2-1 depicts a basic SAN with HBAs.

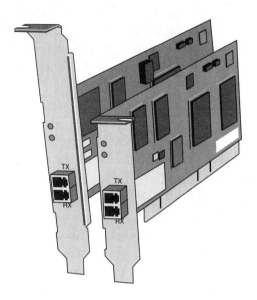

Figure 2-1
Entry-level HBA

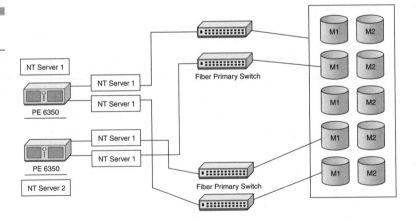

Figure 2-2
Basic SAN

Protocol Basics to SAN Services

Like networking, Fibre Channel uses a layered approach in its architecture. The five layers in the Fibre Channel architecture are FC-0, FC-1, FC-2, FC-3, and FC-4.

FC-0 Layer: Physical Layer

FC-0 defines the basic physical link, including fiber, connectors, and optical/electrical parameters for a variety of data rates.

FC-1 Layer: Transmission Encode/Decode Layer

FC-1 defines transmission protocols, including serial encoding and decoding rules, special characters, timing recovery, and error controls.

FC-2 Layer: Framing and Signaling Layer

FC-2 performs basic signaling and framing functions and defines the transport mechanism for data from upper layers of the stack.

FC-3 Layer: Advanced Features

FC-3 defines generic services available to all nodes attached to a SAN fabric. Some of the services include login server, name sever, management server, and time server, among others, depending on the vendor.

FC-4 Layer: Protocol Mapping

FC-4 specifies mapping rules for several legacy upper-layer protocols and allows Fibre Channel to carry data from other networking protocols and applications. This hierarchical structure is depicted in Table 2-2.

Table 2-2

Hierarchical structure of the five layers in Fibre Channel architecture

Layer	Purpose	Typical Use
FC-4	Upper-level protocol mappings	SCSI III, IP, ATM, HIPPI, VI (a clustering protocol)
FC-3	Generic fabric services	Login server, name server, management server, time server
FC-2	Identifies type of flow control to use for movement of data by sending correct primitives to kick start and stop data transfers.	Dedicated connections with acknowledgment of delivery
		Connectionless with acknowledgment
		Connectionless with no acknowledgment
		Fractional bandwidth for virtual circuits
FC-1	Encode and decoding incoming data	Filtering of characters into data or the correct primitives for transfers
FC-0	Physical media type	Specs for transmitting and receiving signals at different transfer rates

Fibre Channel Topologies

The available options in Fibre Channel topologies are

- Point-to-Point
- *Fibre Channel Arbitrated Loop* (FC-AL)
- Fabric Switched

Point-to-Point

Fibre Channel point-to-point is a simple dedicated connection between two devices. It is used for minimal server and storage configurations. Considered a starter or beginner SAN, point-to-point cabling typically runs directly from one device to another without an intervening hub using a subset of FCPs between the two devices. For additional devices, storage managers can extend the point-to-point cabling scheme. The medium is no longer under the exclusive control of two nodes, so the Arbitrated Loop protocols must be introduced to negotiate access. Figure 2-3 illustrates the point-to-point topology.

Arbitrated Loop

Of the available options, Arbitrated Loop uses twin-axial copper and offers the lowest cost per device; hence it is often the preferred choice. FC-AL is a shared gigabit medium for up to 126 ports connected in a circular daisy chain (one of which may be attached to a switched fabric). Arbitrated Loop is analogous to Token Ring or FDDI. The two communicating nodes possess shared media only for the duration of a transaction; then they yield control to other nodes. Data are transferred from one device to another within the chain. In a typical arbitrated loop with a single initiator and several target devices, the outbound cable from each device becomes the inbound

Figure 2-3
The point-to-point topology

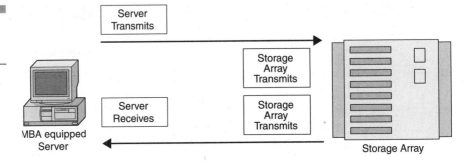

cable for the next device within that loop. FC-AL has become the most widely endorsed open standard for the SAN environment. In addition to high bandwidth and high scalability, FC-AL has the unique ability to support multiple protocols, such as SCSI and IP, over a single physical connection. This enables the SAN infrastructure to function as a server interconnect *and* as a direct interface to storage devices and storage arrays. FC-AL topology is illustrated in Figure 2-4.

Fabric Switched

Switched topology uses switches to establish many concurrent connections between different nodes. Each connection has its own dedicated bandwidth. A switched fabric enables movement of large blocks of data. It is scalable from small to very large enterprise environments and can support high levels of availability. A switched fabric is more costly to purchase but generally has a lower cost of ownership than hubs in any mission-critical environment. Although switched fabrics require some infrastructure implementation, and management software can create some interoperability problems, they offer the greatest connectivity capabilities and the largest total combined throughput. Each device is connected to a switch. Each receives a nonblocking data path to any other connection on that switch. Figure 2-5 illustrates a Fabric Switched topology.

Understanding How a SAN Works

Figure 2-4
The FC-AL topology

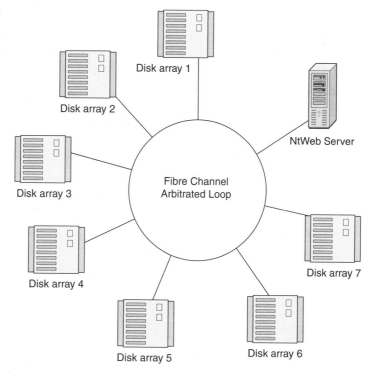

Fibre Channel Architectures by Design

An *architecture* can be defined as the arrangement and selection of hardware to use to facilitate the demand of the storage applications, its traffic patterns, and the customers' degree of obtaining ever-increasing availability. Brocade, a major SAN switch vendor, classifies building blocks of its SANs with four basic design architectures. Attributes of architecture that you should become familiar with are *redundancy* and *resiliency*.

Figure 2-5
A Fabric Switched topology

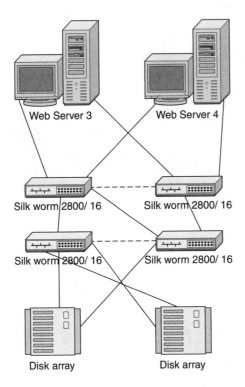

The following defines and illustrates the four basic design architectures (edge switches, core switches, and ports):

8/2/96 Architecture

Explanation
- Eight edge switches
- Two core switches
- Three 96 ports
- A total switch count of 10
- A 3:1 ISL oversubscription rating

Figure 2-6 illustrates the 8/2/96 architecture.

Understanding How a SAN Works

Figure 2-6
The 8/2/96 architecture

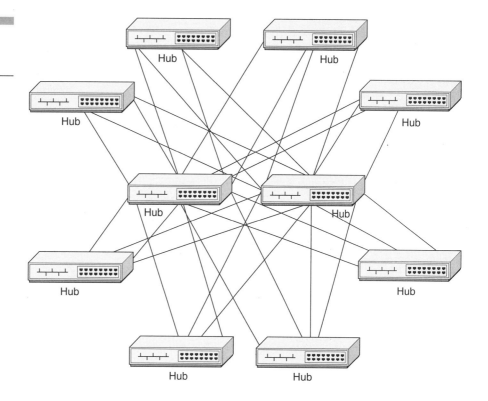

16/4/192 Architecture

Explanation

- Sixteen edge switches
- Four core switches
- 192 ports
- A total switch count of 20
- A 3:1 ISL oversubscription rating

Figure 2-7 illustrates the 16/4/192 architecture.

Figure 2-7
The 16/4/192 architecture

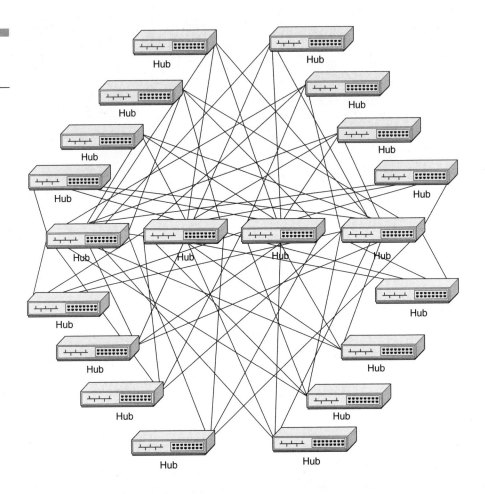

16/2/224 Architecture

Explanation
- Sixteen edge switches
- Two core switches
- 224 ports
- A total switch count of 18
- A 7:1 ISL oversubscription rating

Figure 2-8 illustrates the 16/2/224 architecture.

Understanding How a SAN Works

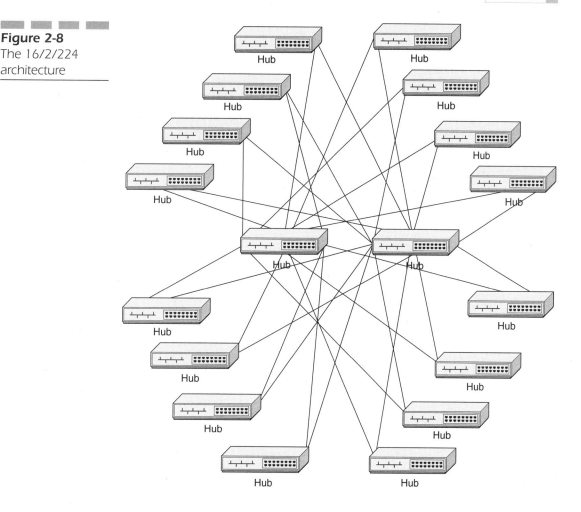

Figure 2-8
The 16/2/224 architecture

12/2/160 Architecture

Explanation
- Twelve edge switches
- Two core switches
- Ninety-six 160 ports
- A total switch count of 14

    Figure 2-9 illustrates the 12/2/160 architecture.

Figure 2-9
The 12/2/160 architecture

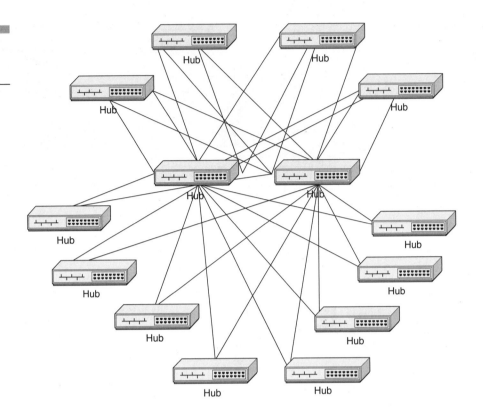

Other Emerging Technologies

While Fibre Channel is the technology of choice for many SANs in today's market, the *Internet Small Computer Systems Interface* (iSCSI), pronounced "iscuzzy," is a technology that uses Ethernet as a medium to connect servers to storage, and it is gaining considerable attention. One reason is because Fibre Channel is a fairly complex technology. As such, most IT staff have limited, if any, Fibre Channel expertise. Therefore, installation of a Fibre Channel SAN can be a bit challenging. When using SCSI protocols mapped onto

Understanding How a SAN Works

the Fibre Channel physical interface, the learning curve typically faced by migration to a new interface can be greatly reduced. Once the software and hardware are installed, using and maintaining a Fibre Channel SAN only requires a moderate level of expertise.

Installation of an iSCSI SAN is considered fairly easy. iSCSI SAN installation leverages the existing IT staff's knowledge of SCSI and Ethernet and embraces the two widely used technologies—SCSI commands for storage and IP protocols for networking. Maintenance and ease of use are considered moderate as well, since familiar Ethernet technology is used to transport data. In terms of performance, Table 2-3 presents a 3-year comparison between Fibre Channel SANs and iSCSI SANs (comparisons are presented in half-duplex).

From this table we can understand why iSCSI merits our attention. Other SAN interfaces that are used include Infiniband, *Enterprise Systems Connection* (ESCON, developed by IBM based on Star topology), SSA (developed by IBM), HIPPI (a dual simplex, point-to-point technology that operates at 800 MBps), or dedicated wide area interconnects such as SONET.

Therefore, while Fibre Channel is used most often, SANs can employ several different types of high-speed interfaces. In fact, many SANs today use a combination of different interfaces. SCSI interfaces are used frequently as subinterfaces between internal components of SAN members (such as between raw storage disks and RAID controllers). Presently, however, Fibre Channel is such an integral part of the SAN solution that we have devoted all of Chapter 5 to discussing it in extensive detail.

Table 2-3

Fibre Channel and ISCSI SANs performance comparisons

	2001	2002	2003
Fibre Channel	200 MBps	200 MBps	1,000 MBps
iSCSI	100 MBps	1,000 MBps	1,000 MBps

Summary

Fibre Channel is the most commonly used technology that powers a SAN. It allows for an active, intelligent interconnection scheme called a *fabric* to connect devices. Fibre Channel, however, is fairly complex, and installation of a Fibre Channel SAN can be challenging. Some companies have opted to use other high-speed interfaces, including iSCSI (which leverages the IT staff's existing knowledge of SCSI and Ethernet), Infiniband, ESCON, or combinations of several different types of interfaces. In Chapter 3, "SAN Setup," we will examine the SAN setup and familiarize ourselves with common SAN terminology.

CHAPTER 3

SAN Setup

Fibre Channel *storage area networks* (SANs) can be designed as both shared-media and switched-access networks. In shared-media SANs, all devices share the same gigabit loop. The problem with this is that as more devices are added, throughput goes down. While this may be acceptable for very small environments, a backbone based on Fibre Channel switches will increase a SAN's aggregate throughput. One or more switches can be used to create the Fibre Channel switching fabric. Accessing the services available from the switching fabric is possible only if the *network interface card* (NIC) of each storage device can connect to the fabric as well as to the operating system and the applications. Basically, the NIC becomes a citizen of the network by logging into the fabric. This function is called simply *fabric login,* and obviously, it is important to use NICs that support fabric login in building a SAN.

Another key issue for devices attached to a SAN is the ability to discover all the devices in the switching fabric. Fibre Channel uses *Simple Name Service* (SNS), a discovery mechanism that learns the address, type, and symbolic name of each device in the switching fabric. SNS information resides in Fibre Channel switches, and NICs and storage controllers request SNS data from the switches. Thus it is advisable to look for Fibre Channel NICs and storage controllers that back SNS.

Combining Switch Access and Shared Media

Some companies are beginning to combine switched access and shared media in designing their SANs. The hard part is trying to determine which parts of the SAN to deploy as switched and which as shared.

In a combination design, *redundant arrays of independent disks* (RAIDs) are attached to Fibre Channel switches, *just a bunch of disks* (JBODs) are connected to Fibre Channel hubs, and a tape drive is linked to a Fibre Channel bridge. For small sites (one or two servers and one or two RAIDs or JBODs), there are three choices:

SAN Setup

- Stick with a SCSI-only model.
- Migrate to a point-to-point Fibre Channel network (with one NIC connected to each storage device).
- Implement a shared-access (hub-based) SAN using a loop topology.

The main considerations are cost and ability to grow. Networks with multiple servers and intelligent arrays housing 500 GB or more of data are candidates for immediate adoption of a switch-based SAN. Consider a midsize company with three servers, multiple RAIDs, and a tape system. A mix of switched attachments to RAIDs, servers with arbitrated loops for multiple JBODs, and a bridge-attached tape system may be appropriate. Moreover, because this SAN is built around a switch, it allows for fault isolation and other switched fabric services. Of course, determining how much of a SAN should be shared and how much switched has to be decided on a case-by-case basis. Principal factors to consider are how mission-critical the data are, the distance requirements, the management of storage devices, the availability and disaster-recovery requirements, business continuity plans, and the ability to manage or cope with configuration management demands.

Let's start with how critical data access is to an organization. In storage setups with parallel SCSI links, for example, the server is the control point. When a server fails, it may take 30 to 90 s for a reset, which could add up to thousands of dollars for a company that provides online billing transactions. In this instance, do not opt for a shared-media SAN because it does not eliminate the reset time; arbitrated loops go through a *loop initialization process* (LIP), causing the reset of all devices. If access to data is a competitive requirement, switched fabrics are the preferred way to go. If storage requirements include any real distances such as across multiple buildings, SCSI probably will not be suitable because there is a length limitation of about 70 m, even with SCSI hubs and repeaters.

As for monitoring the status of devices in multiple buildings, Fibre Channel SANs offers built-in management facilities. Departments using JBODs on loops can be connected to switches on SAN backbones. Switched backbones then create a virtual data center by directly attaching servers and arrays with the loops, giving network managers the management data.

Finally, when it comes to disaster tolerance, a switched fabric is the right choice. Creating redundant data centers 10 km or more apart requires high bandwidth synchronization that only switching offers. SAN setup is discussed in Chapter 4, "Configuring SANs: Dos and Don'ts."

Managing a SAN

As for management of a SAN, it is suggested that you look to use all the tools and systems for your SAN that you use for your *local area network* (LAN) and *wide area network* (WAN). This means that you should look for SAN devices that can be managed via the *Simple Network Management Protocol* (SNMP) or through the Web via the *Hypertext Transfer Protocol* (HTTP). Devices also should support telnet (for remote diagnostics or servicing). Another option for reporting on SCSI devices is *SCSI Enclosure Services* (SES). All of these should furnish detailed information on device status, performance levels, configuration and topology changes, and historical data. Key status and performance information would include throughput and latency metrics. Future requirements will include a commitment to delivering tuning and optimization tools. In *Fiber Channel Arbitrated Loop* (FC-AL) networks, the hub furnishes management information on all devices within the loop. However, the hub cannot report on devices *outside* the loop. Of course, when loops are attached to a switching fabric, remote management and diagnostics are possible for all devices. When deciding on what scheme to use for connecting loops to switches, look for features that make the best use of loop *tenancy*, which is the time a device exclusively occupies the loop for data transfer. Consider devices that collate packets intended for a single target rather than adding the overhead of arbitrating for loop tenancy for each packet to be delivered. More on managing a SAN is presented in Chapter 6, "Case Studies."

Common SAN Terminology and Organizations

Following are SAN-related terminology and organizations that you are likely to encounter. Refer to the Glossary at the end of this book for a complete list of common storage networking-related terms and their definitions.

Directly attached storage (**DAS**)   Previously, all mass storage devices (such as disk drives) were attached directly to the computer or were located inside the same box. Hence there was no need for a term describing this arrangement. Nowadays, since the technologies that enable storage networking have become more diverse, the term *directly attached storage* is used to describe those parts of a wider storage network in which this kind of local connection is still used.

Direct Access File System (**DAFS**)   The DAFS protocol is a new file access protocol specifically designed to take advantage of standard memory-to-memory interconnect technologies such as *Virtual Interface* (VI) architecture. DAFS enables applications to access network interface hardware without operating system intervention and carry out bulk data transfers directly to or from application buffers with minimal *central processing unit* (CPU) overhead.

Fibre Channel Industry Association (**FCIA**)   This is an international organization of manufacturers, systems integrators, developers, systems vendors, industry professionals, and end users.

Host bus adapter (**HBA**)   An input-output (I/O) adapter that connects a host I/O bus to a computer's memory system. HBA is generally used in SCSI contexts.

Just a bunch of disks (**JBODs**)   A term used for a storage enclosure that is supplied with disk drives preintegrated.

The system's integrator can incorporate his or her own choice of RAID controller or just use the JBODs as an economic way to add more disk storage.

***Network-attached storage* (NAS)** A disk array that connects directly to the messaging network via a LAN interface such as Ethernet using common communications protocols. It functions as a storage appliance on the LAN. The NAS has an internal processor and an OS microkernel-based architecture and processes file I/O protocols such as *Server Message Block* (SMB) for NT and *Network File System* (NFS) for UNIX operating systems.

***Redundant array of independent disks* (RAID)** This is a family of techniques for managing multiple disks to deliver desirable cost, data availability, and performance characteristics to host environments.

***SAN-attached storage* (SAS)** This is a term used to refer to storage elements that connect directly to a SAN and provide file, database, block, or other types of data access services to computer systems. SAS elements that provide file access services are commonly called *network-attached storage* (NAS) devices.

***Storage Networking Industry Association* (SNIA)** An umbrella organization we embrace that provides education and services for storage technologies.

Storage subsystem An integrated collection of storage controllers, storage devices, and any related software that provides storage services to one or more computers.

***Storage resource management* (SRM)** Software that keeps track of the use of storage devices in a network and warns system administrators before they run out of disk space.

***Storage Performance Council* (SPC)** An industry standards organization that defines, promotes, and enforces vendor-neutral benchmarks to characterize the performance of storage subsystems.

***Network-attached storage* (NAS)** This is a specialized file server that connects to the network. It uses traditional LAN protocols such as Ethernet and *Transmission Control Protocol / Internet Protocol* (TCP/IP) and processes only file I/O requests such as *Network File System* (NFS) for UNIX and *Server Message Block* (SMB) for DOS/Windows.

Understanding the Difference Between NAS and SAN

NAS is hard-disk storage that is connected to application servers via the network. NAS is set up with its own network address. Since it is not attached to the department computer that is serving applications to a network's workstation users, application programming and files can be served faster because they are not competing for the same resources. The NAS device is attached to a LAN (often an Ethernet network) and assigned an IP address, and file requests are mapped by the main server to the NAS file server. *NAS can be included as part of a SAN.*

A SAN is a collection of networked storage devices that can automatically communicate with each other, a subnetwork that interconnects different kinds of data storage devices as part of the overall network of computing resources for an enterprise. A SAN can be clustered in close proximity to other computing resources, or it can be in a remote location. A SAN supports disk mirroring, backup and restore, archiving and retrieval of archived data, data migration from one storage device to another, and the sharing of data among different servers in a network. SANs may include NAS-enabled devices, but they are not the same thing.

Unlike a SAN, with NAS, users can directly access stored data without server intervention. A SAN, however, will automate management of storage systems, whereas NAS devices do not have this capability. One of the major shortcomings of NAS architecture is that the network on which NAS runs is also used for data access by clients to retrieve data from the file server or communicate with application servers. The data movement between the disk and tape servers also goes over the same LAN. This creates a major network bottleneck when the number of users increases. While NAS works well for documents, file manipulations, and transaction-based applications, it is often not quite so advantageous for database applications because it is file-oriented. For high-bandwidth video applications, NAS slows down because the shared network on NAS gets clogged fast with multiple large files.

Summary

Fibre Channel SANs can be designed as both shared-media and switched-access networks. In shared-media SANs, all devices share the same gigabit loop. Switches will increase a SAN's aggregate throughput. One or more switches can be used to create the Fibre Channel switching fabric. Accessing the services available from the switching fabric is possible only if the NIC of each storage device can connect to the fabric as well as to the operating system and the applications.

Some companies are beginning to combine switched access and shared media in designing their SANs. In a combination design, RAIDs are attached to Fibre Channel switches, JBODs are connected to Fibre Channel hubs, and a tape drive is linked to a Fibre Channel bridge. For small sites (one or two servers and one or two RAIDs or JBODs), there are three choices: (1) stick with an SCSI-only model, (2) migrate to a point-to-point Fibre Channel network (with one NIC connected to each storage device), or (3) implement a shared-access (hub-based) SAN using a loop topology.

As for managing a SAN, you should look for SAN devices that can be managed via SNMP or through the Web via HTTP. Devices also should support telnet (for remote diagnostics or servicing). In Chapter 4 we will discuss the dos and don'ts of configuring a SAN, take a look at migrating from a legacy to a SAN environment, and explore some of the emerging technologies.

CHAPTER 4

Configuring SANs: Dos and Don'ts

Most of us take for granted the storage of data as a technological achievement occurring independent of our recognition and operating solely as a behind-the-scenes process. The closest laypeople come to data storage is use of their floppy drive or in configuration of network tree designations for their particular compartmentalized information. Data storage is much more than this. It is an essential component of daily recognized and unrecognized computer activities. Within various industries, myriad tasks such as telephone communications, banking and *automatic teller machine* (ATM) transactions, airline reservations, radio or television programming, and e-commerce depend on fast, effective data storage. *Storage area networks* (SANs) serve as the preeminent emergent technology of today for bringing islands of information together. As stated in previous chapters, SANs interconnect servers and storage devices at high speeds called *gigabits*. Speedy connection times minimize the need for backup servers. Speedy connections improve interoperability, data management, and security. SANs eliminate network gridlock and are scaled to fit organizational needs. How do legacy systems interact in the SAN methodology, you may ask?

Within the last 10 years, the Internet has grown from thousands of users to millions (53.5 million in 2001). This increase, coupled with an increase in the amount of data generated by users and organizations, has resulted in information aggrandization. Statistics confirm that within the last 5 years, the growth of data storage has quadrupled. Businesses store approximately 50 times the amount of data now as they did 10 years ago. Data storage is expected to continue its increase 20 to 50 times over in the next few years. Internet storage is expected to increase at even faster rates. Growth of information and data storage overwhelms current systems that presently store and manage data. These legacy systems have proven to maintain unique considerations while adapting to the SAN environment.

SAN developments have increased the speed, reliability, and capacity of mass storage technologies, which makes it possible to save limitless quantities of information. Storage compression has heightened nearly 100 percent every one and a half years. Legacy systems have generated isolated units of information, and computer technology has changed dramatically over the past decade, delivering a multitude of variables for data storage. These variables are not compliant with one another. Computers housed within the same facilities often are elementally different and incompatible. Yet each system maintains a unique data storage capacity and functionality. Variables such as these result in data that are not easily accessible across an enterprise.

The connections through which these systems exchange data do not transmit mass quantities of information well. They do not process at quick rates; typically, they do so at a snail's pace. As exponential information and data storage growth continues, the difficulties in managing, protecting, manipulating, and/or sharing data mount statistically higher each year. The implementation of a SAN changes this trend.

Configuring from Existing Legacy Systems

Seventy percent of mission-critical data are stored on mainframes, and until recently, the SAN community has all but ignored these legacy systems. Figure 4-1 provides an overview of an old enterprise or legacy system.

The clients, or end users, work from their workstations, where stored data are backed up over the *local area network* (LAN). The LAN maintains UNIX, WINNT, Netware, or other legacy server types. Islands of *Small Computer System Interface* (SCSI) disks process and provide first-level storage of data that are then backed up to a tape library system. The legacy system as represented in Figure 4-1 requires scalability, higher capacity, and increased storage capacities, and a SAN will provide this through LAN-free backups. LAN-free backups enable the SAN to share one of the most expensive components of the backup and recovery system—the tape or optical library and drives within it. Figure 4-2 depicts the evolution of a centralized backup.

There was a time when backups were completed to local attached tape drives of the legacy system. This method worked when data centers remained small and each server could fit on a single tape. When management of dozens or hundreds of tapes became too difficult when servers no longer fit on tape, data centers began to implement backup software that allowed them to use a central

Figure 4-1
Old enterprise or legacy system

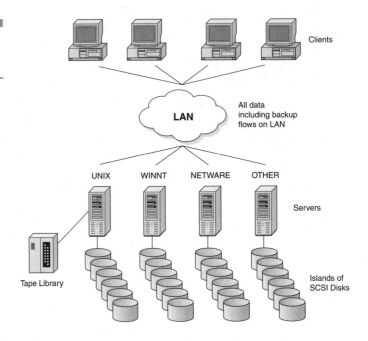

Figure 4-2
The evolution of a centralized backup

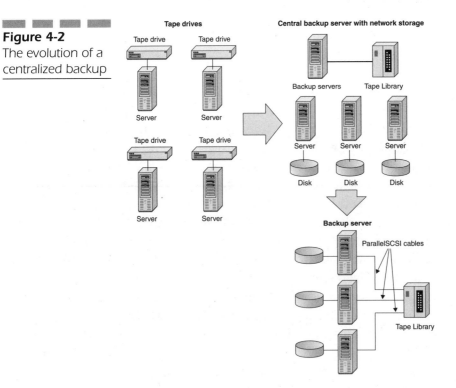

backup server and the backed-up servers across the LAN. Servers are referred to as *clients* by backup systems.

This methodology worked as long as the LAN could support network traffic in addition to its backups. Even with state-of-the-art LANs, you encountered individual backup clients that were too big to support on an enterprise-wide basis across the LAN. Large amounts of system resources were required on backup servers and clients when backing up these large amounts of data across the LAN. Software backup companies realized that this was going to occur and began to offer support for remote devices. You could decentralize backups by placing tape drives within each backup client. Each client then would be told when and what to back up by central backup servers, and the data would be transferred to locally attached tape drives afterwards. Most major software vendors allowed this to be done through a tape library. You can see the evolution of backups in Figure 4-2. There are connections with one or more tape drives from tape libraries to each backup client maintaining requirements. The physical movement of media within libraries was managed centrally, usually by backup servers.

The configuration depicted here is referred to as *library sharing*. The library is being shared, but the drives are *not* being shared. When people talk about LAN-free backups, they refer to drive sharing. In drive sharing, multiple hosts maintain shared access to an individual tape drive(s). Library sharing requires dedicated tape drives to back up clients where connected. Tape drives within a shared library remain unused most of the time.

If you were to take three servers, as an example, with 1.5 TB of data each, where 5 percent of the data would change daily, resulting in 75 GB of backup data per day per host, and those backups must be completed in 8-hour windows, the host would be backed up in the same window at aggregate transfer rates of 54 MBps for full backups. If you assume that each tape drive is capable of 15 MBps and that each host requires four tape drives to complete the full backup within one night, you would require four tape drives for each server, resulting in a configuration that would look like Figure 4-2. This configuration allows servers to complete backups in the backup window. The many tapes drives allow completion of incremental backups (75 GB) in approximately 20 minutes.

Figure 4-3
LAN-free backups

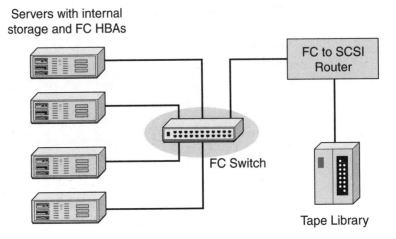

If you take these same servers and connect them along with the tape library to a SAN, as illustrated in Figure 4-3, your result would be far different. Each host has a switched fabric SAN via a single 100 MB Fibre Channel connection, along with tape drive connections (if tape drives are supported to Fibre Channel natively), so each can connect to the SAN via switch. If the tape drives connect via standard parallel SCSI cables, you could connect five tape drives to the SAN via Fibre Channel routers. Using four-to-one modeling for Figure 4-3 enables tape drive connects to the SAN via a single Fibre Channel, enhancing performance dramatically.

SAN Configuration

System Requirements and Analysis

Prior to building a SAN, you need to analyze your system requirements. Ask yourself the following questions:

Configuring SANs: Dos and Don'ts

- What problems are being resolved through implementation of the SAN environment?
- What is the technical requirement for the business operation?
- What is the broad case of the business environment?
- What are the corporate and/or project goals in SAN implementation?
- Is there a timeline? Is there a cost/benefit analysis that needs to be recognized?

Figure 4-4 illustrates the SAN life cycle.

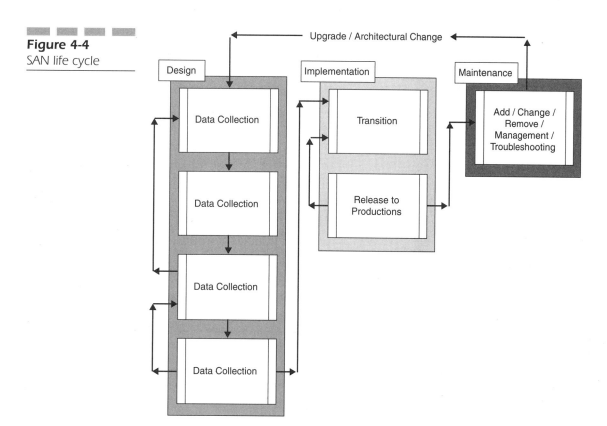

Figure 4-4
SAN life cycle

SAN Architecture

When configuring a SAN, be sure to develop architecture that meets your system requirements and analysis. SAN architecture development requires good management that recognizes operating and performance considerations, successful data collection, and implementations of phase/cycle development procedures. A detailed SAN architecture includes fabric topologies of all related fabrics, storage vendors, SAN-enabled applications being used, and considerations that affect your overall SAN solution. Contingent on the definition and development of requirements and needs, your SAN architecture can be either timely or a short-lived process. It is good to recognize that this should be a precise developmental phase.

Building a Testbed

You need to develop a prototype of the SAN architecture solution that you established and then test it so that functionality will be fully adopted. Testing should be completed using nonproduction system analysis methodologies. This component of the SAN life cycle may involve readdressing the SAN architecture if any inconsistencies are found during testing. Testing your SAN architecture may or may not be a totally plausible concept. With large systems, it may not be an easily accommodated task to test all aspects of a SAN configuration; however, often there are documents that relate certifications of configuration from other vendors that expensed the testing for you. Using these certifications and remembering your specific SAN architecture perhaps will alleviate the requirement of testing before implementation. Still, in all, compliance testing is required because of the nuances that may arise in timing or interactivity or because you could experience device failures. A test plan is required prior to releasing your SAN architecture for production. Take your time to test and access your SAN environment. Noting any potential weaknesses will be worth its weight in gold if they are resolved during testing and prior to release.

Legacy-to-SAN Environment

Figure 4-5 shows a new enterprise system and/or legacy-to-SAN environment. This figure illustrates how storage pools interact in the old enterprise system. The SAN architecture is ideal to handle large amounts of data requiring high accessibility. The most common communication infrastructure for a SAN is Fibre Channel (a serial switched technology that operates in full-duplex mode with maximum data rates of 100 MBps in each direction). Originally, Fibre Channel was designed to compete with and possibly replace Fast Ethernet. Fibre Channel is now nearly synonymous with data storage. Fibre Channel supports distances of up to 10 km, and this is an improvement over SCSI's 6 m limitations. Fibre Channel runs over fiberoptic and copper cabling, although copper cabling does restrict the cable length to 30 m. Fibre Channel supports multiple protocols, including SCSI, *Internet Protocol* (IP), and *Asynchronous Transfer Mode* (ATM) *Adaptation Layer 5* (AAL5).

In theory, SANs enable users to access information from any platform (this is important in mixed Windows-UNIX installations that are so very common), share storage devices and files, and readily

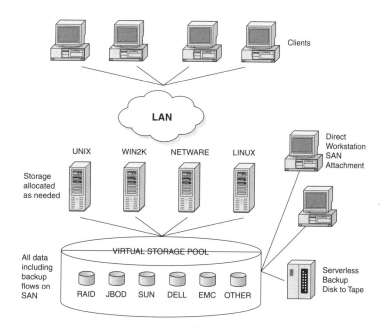

Figure 4-5
New enterprise system (legacy-to-SAN environment)

increase storage capacity to meet growing needs. This all takes place while reducing the impediments of clogged LANs and *wide area networks* (WANs). SANs are entire storage hierarchies with *network area storage* (NAS), disks, tapes, tape libraries, and other archiving devices.

Is there a downside to SAN implementation? There are always drawbacks to system upgrades. One problem is that a SAN introduces another network. Like LANs or WANs, SANs require management software, utilities, operating system supports, security, maintenance, and training. These added complexities easily could outweigh the advantages; sometimes doubling network staff workload is not feasible when there is already plenty to do and no one available for hire. SANs represent major infrastructure changes that require compelling justifications, and they require higher-bandwidth cabling and are not currently extensible over long distances. Are there solutions to these drawbacks? Yes, there are.

One solution calls for the use of set protocols throughout all networks—to run storage networks over IP. There are several alternatives, including storage over Gigabit Ethernet, *Internet SCSI* (iSCSI), and Fibre Channel over IP. Figure 4-6 exhibits a Fibre Channel solution. Gigabit Ethernet looks attractive on the surface

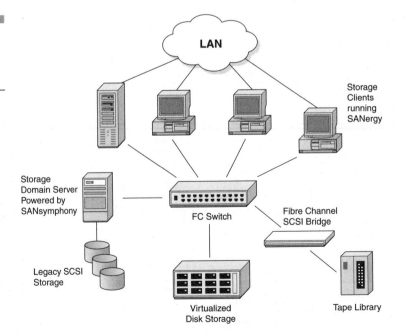

Figure 4-6
Fibre Channel switches in a SAN system

Configuring SANs: Dos and Don'ts

because it avoids duplication of training, test equipment, and software. The iSCSI/Fibre Channel alternative shown, though, preserves current investments of storage devices, software, and management methods. Still another possibility includes the use of emerging standards to provide migration paths from all past approaches.

Solutions for wide-spectrum enterprise storage and management needs include

- Data backup and recovery
- Backup/restore efficiency
- Consolidation of storage resources
- Improved use of storage resources
- Enhanced business through a modernized storage infrastructure
- Storage management solutions
- Data archival
- Improved flexibility and connectivity of storage resources
- Mass storage solutions
- Devices that can be attached to the SAN
- Communication patterns including random *input/output* (I/O), streaming accesses such as video, and I/O-intensive database accesses
- Performance characteristics: reads, writes, max/min/typical throughputs

Storage solutions are delivered to customers in one of three ways:

1. Building custom solutions tailored to the client's specific needs:
   - Backup and recovery
   - Archival
   - Data management
   - Clustering and replication
   - Disaster management and recovery
   - Storage infrastructure enhancements

2. Provision of integrated product solutions addressing specific application requirements
3. Developing and introducing proprietary solutions, including
   - Complete turnkey solutions for content management and distribution of resources such as video streaming products over broadband networks
   - Solutions for geospatial data management and storage
4. Providing storage system management and consulting services to clients, including
   - Consultations to assess current storage infrastructure capabilities
   - Determinations of future requirements and technology choices
   - Architecture of engineering solutions
   - Installation and integration solution components overview and analysis
   - Solutions management

Storage Performance Survey

The first step to developing quality storage solutions is to perform a discovery evaluation of your current storage performance, data availability, and future storage needs. Engineers can, with the assistance of staff, perform and execute rigorous site surveys to develop comprehensive statements regarding current and future data management requirements. Then the requirements must be mapped against the business mission or objectives. Storage performance surveys may include

- Written assessment of the current *information technology* (IT) storage infrastructure
- Network and data flow diagrams
- Storage use and optimization analysis

Storage Architecture and Design

The key to powerful networked storage design is in the quality of the design and the response to your requirements. Well-organized knowledge bases are necessary to design enterprise-class information and storage infrastructures. Talented, experienced storage engineers will provide custom data management solutions to meet broad continuums of challenges and deliverables that should be supported by detailed, documented strategies, including diagrams and implementation plans for data management. When appropriate, cost/benefit plans showing payback periods and returns on investment, including optional lease analyses for all proposed products and services, can be requested. Storage architecture and design planning should include

- Written recommendations for the storage infrastructure
- Complete design specifications
- Installation and integration

Whether you hire certified engineers through consultancy or designate in-house engineering teams to implement and configure your SAN, you will need to have them do the following:

- Install the hardware load.
- Configure software for all backups.
- Archive.
- Perform media management.
- Provide off-site vaulting, if necessary.
- Implement verification and testing of backups, archivals, and restores.
- Develop operating policies and procedures.
- Recommend changes where appropriate.
- Provide detailed information on installation and integration, implementation, storage management software installation,

switches, hub and adapter integration, *redundant arrays of independent disks* (RAID) reconfigurations, server consolidation, and so on.

These solutions link multiple and disparate storage systems and servers over secure high-speed storage networks dedicated to data management. Integration of these solutions enables storage efficiency across tape libraries and disk arrays with optimal shared resource pooling. These solutions contribute to

- Improved data availability
- Decreased administrative complexity and costs
- Improved asset utilization
- Reduced downtime
- Point solutions (backup and recovery solutions, archival systems, automated storage management data environments)
- SAN-ready architecture (implements storage infrastructure components that enable you to take advantage of new technologies and scalability features as you choose)
- True SANs (fully networked Fibre-Channel-based SAN with maximum scalability and advanced data management capability, the most powerful storage infrastructure for the greatest operational cost savings)
- Client site outsourcing (proactive on-site storage systems management)

Recognizing IT environments and requirements varies widely. A comprehensive range of connectivity and storage options can meet your requirements. Heterogeneous multihost access to centrally managed pools of storage and operating system support may include use of Solaris, Hewlett Packard/UNIX, AIX, Irix, Microsoft Windows NT/2000, and/or Linux products, among others.

Return on Investments

With any business, it is important to first quantify the economic benefit based on cost/benefit analysis and/or the *return on investment* (ROI) your organization will receive. Preparing an ROI report for your SAN project, as illustrated in Figure 4-7, shows how you will benefit from implementation of the SAN environment. The ROI report will show the return on capital investments and save you money in time, management, and other efficiencies through foreknowledge or needs and finances versus requirements.

Through an interview process, you make a list of all the equipment that you will require and determine what components are specific to your SAN. If you need to buy additional storage arrays, whether or not you implement a SAN, this should be included as an expense in the analysis. If your SAN will prevent you from needing

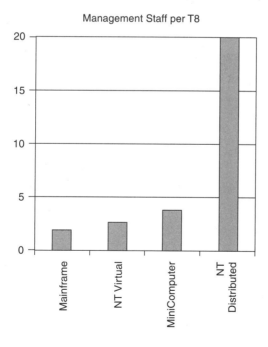

Figure 4-7
Management ROI sample

to purchase an array, this savings in cost will go to the benefit or credit side of your analysis. You should include testing hardware that may not be used elsewhere in your SAN development as an expenditure.

When you account for staff time spent on the project, make certain that you only charge for time spent beyond what would be spent in not building the SAN. If you anticipate staff timesaving in the final scheme, apply this to the benefit side of the analysis. ROI analysis will be continuing as a living document and will be updated as the SAN project develops. A well-prepared ROI report can attest through analysis to the need for proposed implementations. Follow these steps to produce an ROI analysis based on SAN solutions that answer specific needs:

1. Theme or scenario. Decide on the purpose of your SAN implementation, such as
   - Server or storage consolidation
   - Infrastructure usage improvement
   - Economic justifications, for example, savings
   - Storage and server resources utilization
   - High-availability clustering
   - Mission-critical applications availability improvements
   - Improved data integrity

2. Components. SAN deployments focus on affected servers. Servers are grouped according to the applications they run or functional support areas. Examples of application groupings include
   - Web servers
   - Files
   - Print servers
   - Messaging servers
   - Database servers
   - Application servers

Configuring SANs: Dos and Don'ts

Functional support servers might include
- Financial
- Personnel systems
- Engineering applications
- Amount of attached disk storage
- Storage growth rates
- Storage space reserved for growth (headroom)
- Availability requirements
- Server downtime
- Associated downtime cost
- Server hardware and software costs
- Maintenance costs
- Administration required to keep servers up and running

3. Identify SAN benefits.
   - Server and storage consolidation
   - Efficient use of server and storage resources
   - Improved staff productivity
   - Lower platform costs
   - Better use of the infrastructure

4. Identify SAN costs. Identify new components required for building and maintaining the SAN:
   - Software licenses
   - Switches
   - Fibre Channel *host bus adapters* (HBAs)
   - Optical cables
   - Service costs in deployment

5. Calculate your ROI. There are a number of standard ROI calculations commonly used today, such as
   - Net present value (in dollars)

- Internal rate of return (as a percentage)
- Payback period (in months)

These are defined as

- *Net present value* (NPV). A method to evaluate investments in which net present value of all cash flow is calculated using a given discount rate.
- *Internal rate of return* (IRR). A discount rate where the present value of the future cash flow of investment equals the cost of investment.
- Payback period. Length of time needed to recoup the cost of capital investment on a nondiscounted basis.

Establishing and calculating an ROI enables you to remain focused on infrastructure-based problems and provides reduced deployment risks through phased SAN deployment. This helps keep investments limited to solutions and creates investment bases for the future. Initial investments will improve the ROI when other scenario investments are reduced.

You should have the following detailed results from an interview process that define what the SAN needs to accomplish for you:

- Technical requirements document?
- Operating system to be installed?
- Patch or service pack level?
- Whether fabric HBA/controller drivers are available, and if so, are they well tested?
- How many HBAs will it have?
- If there is more than one HBA, what software will be used to provide failover or performance enhancements of multiple paths?
- Private loop, public loop, or fabric connection?
- Applications that will run host databases, e-mail, data replication, or file sharing?
- Storage requirements?
- Will storage requirements change over time?

Configuring SANs: Dos and Don'ts

- Physical dimensions? How heavy will it be?
- Does it require a rack mount? A rack kit? A shelf set?
- Management console? If so, what type is it?
- Traditional *keyboard/video/mouse* (KVM) combo?
- Serial connection, such as a TTY?
- Do interfaces exist? Do they need to be purchased?
- Make, model, and version information?
- How many Ethernet interfaces?
- What temperature range for operation?
- Telephone line requirements?
- Node location?
- Timeline for implementation?
- Required purchases?
- SAN design?

Answers to these questions will be used throughout the life of your SAN. The timeline will be the framework of activities for the SAN life cycle. These documents will be used in approval processes during implementation and maintenance phases as SAN deliverables. If major changes to your SAN environment are necessary, your SAN life cycle will be repeated, and another set of documentation will need to be produced.

Interoperability Issues

Currently, interoperability is limited. Equipment from different manufacturers often will not work in an interoperative fashion despite common software and hardware interfaces. Third-party testing has become necessary to ensure interoperability. The Fibre Channel market is widely accepted as the most innovative SAN development solution at this time. Integrated software companies play a major role in the development of SANs. System managers and administrators need to recognize that this is a fact.

E-commerce is often associated with UNIX and Windows NT systems. What is overlooked is the fact that mainframes are integral to the e-commerce supply chain. Telephone companies, shipping companies, credit card companies, and banks all manage multiple mainframes supporting e-commerce. Mainframes are involved in online transactions before the customer clicks on a buy icon and until the product arrives on the customer's doorstep. If *data* is to be synonymous with *business,* then interconnecting open-system data storage with mainframe data storage is a primary role in the new economy.

Interconnecting open-system and mainframe storage is not easy. Data structures, protocols, and architectures vary. Both environments are double-edged swords in relation to interoperability. Their strengths are often their weaknesses in the SAN environment. The nature of an open system is that it is open. The premise of the open-system environment is that multiple vendors can build compatible products to interoperate with one another and allow customers to use common sets of applications and devices. Vendors may or may not develop compatibility products contingent on the market. It is staggering to think about the scope of open-system environments, where chips, processors, boards, interface cards, communications protocols, cabling, operating systems, infrastructure devices, and applications are essential. With thousands of vendors who develop, innovate, and deliver these products to the market, it is no wonder that open systems can be unreliable and unstable.

Mainframes, on the other hand, are closed and proprietary by design. The mainframe environment is stable and reliable. Few vendors seem to control everything end to end. The mainframe world achieves superior reliability and performance because the operating systems, communications protocols, hardware, and system components are tightly integrated.

What is done today to bridge the data storage gap between open-system and mainframe storage? In order to connect these two disparate worlds, companies have engineered around inefficiencies. There are technologies that enable data to be moved between open-system and mainframe storage. They do not always provide advanced functionality, intelligence, or management features. They are most always prohibitively slow.

There are various ways mainframe and open-system data sets can be accessed today. Over standard IP networks, data may be transferred by all standard means including *File Transfer Protocol* (FTP) and *Network File System* (NFS). Some vendors provide middleware to move and transform data between the two environments. There are devices that support SCSI, Fibre Channel, and *Enterprise Systems Connection* (ESCON) interfaces. Most of these products have no intelligence and provide only physical connectivities. Use of storage subsystems in support of SCSI, Fibre Channel, and ESCON interfaces is another solution. Although a storage subsystem has software to perform data transformation, the system running the software is not integrated into the storage subsystem; it is actually a separate process. This causes latency and limitations regarding system functionality and integrity. Storage subsystems are vendor-dependent, which restricts your options.

There are major problems with all these solutions. Each one requires host-based software to perform tasks. Host-based software solutions have proven to be slow and cumbersome; they have a severely negative impact on overall business productivity. Hosts are production systems, and additional processing affects their overall performance. Moving large amounts of data becomes impossible. Most of these solutions do not provide online presence in their data paths. To share data, the mainframe needs an active open new data path. This is done on a point-in-time basis and not continuously, which means that data are redundant on storage.

Companies have moved today from gigabytes of stored data to terabytes of stored data. Predictions are that stored data soon will enter the petabyte range. The solutions mentioned previously offer very little intelligence, scalability, or robustness. They may achieve raw file transfer, but to move volumes of data, multiple simultaneous replications, online backups, and so on between platforms does not exist.

SAN Gateways

Data are essential for businesses, and controlling data is necessary. Open-system data and mainframe data have sum and substance differences. The effective way to connect these environments is through manipulation of data via SAN gateways. Balance is essential to ensure intelligence. Performance is standard in deployment.

Performance without intelligence offers limited solutions. An intelligent SAN gateway sits in the middle of the data path and interconnects with mainframe and open-system storage at interface and data levels. The SAN should be vendor-independent, and it should support multiple platforms.

A shift in the dynamics of open-system and mainframe architectures creates bandwidth deployment issues. Open systems use few "fat" pipes, and mainframes use many "thin" pipes. On an open system, Fibre Channel can move data at a bandwidth of 1 Gb. ESCON supports approximately 17 Mb of bandwidth; this creates bottlenecks. An intelligent SAN gateway addresses these issues by funneling ESCON channels to create fatter pipes. This solution effectively takes ESCON channels to the creation of virtual channels and provides higher bandwidths. A SAN gateway is scalable and has the ability to add port and as much bandwidth as necessary.

Fibre Connection (FICON) is available for mainframe storage. FICON provides benefits that include increased bandwidth. There are some major issues that FICON must resolve before full deployment is possible. Existing mainframe systems cannot upgrade to FICON, so companies put bridges in to connect ESCON and FICON. New mainframes have FICON channels installed automatically, but at this time there are no control units to support new interconnect technologies, which is unfortunate. SAN gateways require existing infrastructure support in adaptation to new technologies as they emerge.

Responsibility needs to be removed from the host and the storage subsystems. It should be put on outboard systems. Proposed SAN gateways capable of moving large quantities of data and of acting on them in real time are necessary. To achieve this without affecting performance requires a great deal of energy; it can be achieved on a per-port basis with RISC processors.

SAN placement is a crucial issue. The SAN should sit in the middle of the data path to intercept and interpret data as they transverse the open-system and mainframe environments. Placing the SAN gateway anywhere else will limit its ability to access and control data, reducing effectiveness. The provision of intelligence is essential so that data can be transmitted rapidly between mainframe and open-system storage. The platform will embed intelligence, working through lightweight agents that reside on various hosts and storage systems. Such devices will extend beyond basic file transfers and raw data conversions. The SAN should perform advanced functions and provide volume transfers, volume remapping, online backups, point-in-time copy, snapshot copy, asynchronous and synchronous mirroring, and other applications.

Heterogeneous enterprise SANs will never exist unless interconnecting mainframe storage is resolved. Basic connectivity and applications are insufficient to meet today's requirements. Intelligent, robust, and powerful solutions that solve the interoperability, performance, and scalability issues found in today's SANs are necessary.

SANs serve the purposes of global enterprises, network service providers, and emerging storage service providers. Users tap into services and storage as necessary and pay for what they require, with no long-term capital investment necessary. Whether upgrading existing data storage systems, designing a SAN architecture, or managing special projects onsite, securing qualified engineers can help you find appropriate solutions to meet the business needs of your particular organization.

Project Management

Project management services that ensure efficient turnkey integration processing with an understanding of technical interdependencies of various storage components and experiences with equipment manufacturers who may deliver critical components of your solution are essential. Comprehensive project methodologies that are flexible and responsive to your particular requirements also are necessary. You should have

- SAN project management
- SAN project executive
- SAN maintenance and support

Examples of Successful SAN Transitions

Before turning to a SAN, a content company relied on combinations of *direct-attached storage* (DAS) and *network-attached storage* (NAS) to house databases that served as their lifeblood. This content company outfitted servers with dedicated storage devices connected via SCSI cables (Figure 4-8). On the surface, this appears to be an appropriate and scalable solution; if project management had more data to add, it simply would add a new server and a new storage device.

As content databases grew by 300 percent, the company ran up against limitations of SCSI-based DAS, including single points of failure, limited numbers of devices on the SCSI bus, and cluttered physical configurations that became difficult to manage. Provisioning new storage took an average of a week and a half . . . far too long to match the company's growth spurts. As you can imagine, this business model demanded instant access to huge pools of data that had to be correlated and served in real time. Without access, this content database provider would have no business. Its problem was running up against the limitations of the storage platform it had chosen originally as its solution. Further limitations could be the length of the SCSI chain and how many devices could be maintained on the single SCSI bus.

This company turned to a SAN to boost availability and performance of content databases (Figure 4-9). The SAN included a 1.8 TB 4500 array and two 16-port 2800 Fibre Channel switches that easily handled millions of daily I/O transactions. Data became available from any server and provided improved reliability and redundancy.

Figure 4-8
Content provider DAS

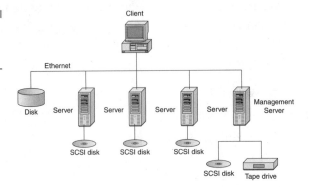

Configuring SANs: Dos and Don'ts

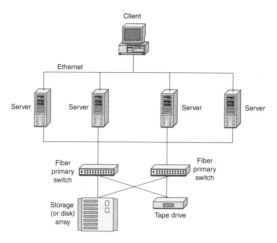

Figure 4-9
Content-provider SAN solution

The SAN slashed configuration time to about 3 days. In addition, the content company found that it could separate client, backup, and shared-file-system network traffic, significantly reducing backup times.

Many organizations are looking into SANs as a way to control bandwidth utilization on their networks while ensuring good quality of service for *enterprise resource planning* (ERP), *customer relationship management* (CRM), data warehousing, and other data-intensive applications. These applications often depend on multiple databases as well as other data sources, so there are no easy rules of thumb to decide when to use SANs, NAS, or DAS.

There are, however, useful questions to be answered about the data needs of your applications. The location, volume, and frequency of the data being moved are primary concerns. If you understand these factors and the related impacts of storage management and staff considerations, you will be able to develop an appropriate storage solution.

Peer Model

The peer model uses a single control plane incorporating an administrative domain that includes a core optical network and related edge devices. The edge devices see the core's topology. They are scalable; a point-to-point mesh is used for data forwarding. Routing protocol information is passed between each device adjacent to its "mother" photonic switch rather than to the other edge devices.

Most carriers use different approaches for different applications depending on their individual network topographies and services supported; hybrid models most likely will be popular. In this scenario, edge devices act as peers to core networks, sharing common control planes with the core. Simultaneously, other edge devices control overlay fashion with their own control plane and interface with the core through a user-network interface. Peer model functionality, as depicted in Figure 4-10, encompasses the requirement of the overlay model so that one suite of control plane protocols supports both models. A carrier can adopt the peer model and go either way, as its business model dictates.

Dense Wavelength Division Multiplexing

Dense Wavelength Division Multiplexing (DWDM) is a fiberoptic transmission technique that employs light wavelengths to transmit data parallel by bit or serially by character. DWDM system scalability is important when enabling service providers to accommodate consumer demand for ever-increasing amounts of bandwidth. DWDM is discussed as a crucial component of optical networks that enables the transmission of e-mail, video, multimedia, data, and voice—carried on IP, ATM, and *Synchronous Optical Network / Synchronous Digital Hierarchy* (SONET/SDH).

The Bandwidth Crisis

As Internet connections worldwide double each year, more users are upgrading to gain faster access. It seems to be a great time for Inter-

Figure 4-10
Peer model

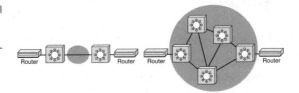

Configuring SANs: Dos and Don'ts

net users and service carriers. Users who subscribe to faster access will enjoy speedier Internet experiences, and service carriers can cash in on the increased user base. Unfortunately, reality is far from ideal, since users are not getting the bandwidth advertised by providers. The Internet carriers are having a hard time balancing between upgrading infrastructure to meet Internet traffic growth and keeping shareholders happy by making profits.

Not only does the bandwidth crisis apply to the Internet sector, but it also applies to voice, fax, and data traffic. Data traffic is gaining in importance over all the other sectors. Internet traffic has been growing exponentially from 1990 to 1997 compared with traditional telecommunications traffic of 10 to 17 percent in the same period. This is illustrated in Figure 4-11.

There are various media such as coaxial cable, twisted pair, optical fiber, radiofrequency, satellites, and microwave wireless, but optical fiber is the only medium with enough bandwidth to handle the rapidly growing high-capacity backbone of the data storage environment in the world today. The most common fiber-optic cable is OC-48, with a transmission rate of 2.488 Gbps. The deployment of 10 Gbps in OC-192 is becoming more feasible. From a user standpoint, 2.5 or 10 Gbps may sound like enormously high bandwidth, but 2.5 Gbps can support approximately 1,600 *digital subscriber line* (DSL)

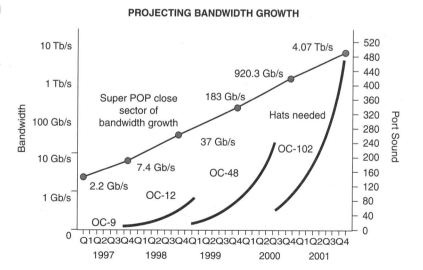

Figure 4-11
Projecting bandwidth growth

users connecting at 1.544 Mbps. This may be fine for a local carrier, but the major network access points around the United States will have trouble keeping up if DSL and other broadband services are as widespread as 56 kbps modems are today. The largest access points —MAE-East (Washington, DC) and MAE-West (San Jose, CA)—have combined traffic that has more than quadrupled in each of the 2 years since 1995. Combined traffic reached about 1.5 Gbps on average in mid-1997. This is an equivalent of about 12,000 phone calls, and this figure indicates that the Internet is insufferably slow because there had to be more Internet traffic than the equivalent 12,000 phone calls around the United States in 1997.

Possible Solutions to the Bandwidth Crisis

There are various solutions to ease the bandwidth crisis. Carriers can install more fiber-optic cable into their infrastructures. For long-distance carriers whose networks exceed 10,000+ miles of cable across the United States, installing more fiber-optic cable is not logistically feasible. Even for small-distance carriers (300 km or less), laying extra cable in a metropolitan area may take months to achieve, requiring government paperwork and clearances prior to beginning construction. In either case, carriers are turning to DWDM as a SAN solution. By combining multiple wavelengths, each representing separate data channels, the same fiber-optic cable suddenly has the bandwidth capacity of multiple cables. An additional benefit of DWDM is that repeaters commonly used in *Time Division Multiplexing* (TDM) networks, such as SONET, are replaced by optical amplifiers. Unlike repeaters, optical amplifiers can amplify multiplexed signals without demultiplexing first. This reduces intermittent bottlenecking. Optical amplifiers also can be placed farther apart than repeaters.

With TDM, bandwidth is inversely proportional to pulse width twice. In other words, a higher bandwidth in a TDM system will create a shorter pulse width, equaling a higher frequency and making it more susceptible to fiber dispersion. As the bandwidth requirement increases in the future, researchers are faced with the dispersion issue if TDM is to be competitive. DWDM increases its

Configuring SANs: Dos and Don'ts

bandwidth by adding extra wavelengths and has a theoretical bandwidth of 5,000 GHz in the 1,550 nm region.

Notice that in applications where recoverability is essential, SONET (TDM) will still be the first choice because of SONET's 50-ms recoverability. In the case of a path interruption, this is something DWDM cannot achieve today by itself.

What Is DWDM?

DWDM is a technology that transmits multiple data signals using different wavelengths of light through a single fiber. As illustrated in Figure 4-12, incoming optical signals are assigned to specific frequencies within a designated frequency band. The capacity of the fiber is increased when these signals are multiplexed out onto one fiber.

The best way to describe DWDM technology is by readdressing the earlier analogy of a car traveling on a highway. The analogy compared the SAN environment and movement of data to one fiber traveling to a multilane highway using one lane. It stated that traditional systems use one lane of the highway. Moving the cars, or signals, faster within the one lane increases the capacity of the highway. DWDM technology accesses the unused lanes, increasing the number of wavelengths on the embedded fiber base and fully using the untapped capacity within the fiber.

Figure 4-12
DWDM

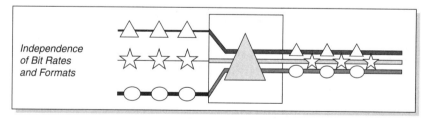

- Merges *optical* traffic onto one common fiber
- Allows high flexibility in expanding bandwidth
- Reduces costly mux/demux function, reuses *existing* optical signals
- Individual channels use original OAM&P

DWTM = Dense WDM

DWDM is able to reach transmission capabilities four to eight times faster than traditional systems. This high-speed, high-volume transmission is made possible by the technology within the optical amplifier. An optical amplifier is a section of fiberoptic cable that has been doped with erbium to amplify the optical signal. *Erbium-doped fiber amplifiers* (EDFAs) have two advantages: they increase optical signals, and they do not have to regenerate the signals to boost strength. A typical optical signal must be regenerated every 100 km. This is accomplished by converting an optical signal to an electrical signal and then back to an optical signal for its retransmission. EDFAs lengthen the distance of transmissions to more than 300 km before regeneration is required.

Amplifiers should adjust automatically when a channel is added or removed. This helps achieve optimal system performance. Degradation in performance through self-phase modulation may occur if there is only one channel on the system with high power, so optimal performance is important. The reverse is true: If there is too little power, there will not be enough gain from the amplifier. Questions have been raised as to which types of optical amplifiers enhance system performance most between fluoride- and silica-based fiber amplifiers. Silica-based optical amplifiers with filters and fluoride-based optical amplifiers both exhibit comparable performance ranges between 1,530 to 1,565, but fluoride-based amplifiers are more expensive, and the long-term reliability of fluoride-based amplifiers has yet to be proven.

Is DWDM Flexible?

DWDM is protocol- and bit-rate-independent, so data signals such as ATM, SONET, and IP can be transmitted through the same stream regardless of speed differences. Each individual protocol remains intact during transmission processes because there is no optical-electrical-optical conversion with DWDM. The signals are never terminated in the optical layer, which allows independence of bit rate and protocols. DWDM technology can be integrated easily with the existing equipment in the network. This gives service providers flexibility to expand capacities within any portion of their networks. No other technologies allow this. Service providers are able to partition dedicated wavelengths for customers who would like to lease one wavelength instead of an entire fiber.

Service providers may begin to increase the capacity of TDM systems currently connected to their networks. This is so because OC-48 terminal technology and related *operations support systems* (OSSs) match DWDM systems. OC-192 systems may be added later to expand the capacity of the current DWDM infrastructure to at least 40 Gbps.

Is DWDM Expandable?

Analysts estimate that annual growth in voice traffic proceeds at a steady rate of between 5 and 10 percent; data traffic is forging ahead at a rate of 35 percent per year in North America. Numbers such as this show that the network technology in current use already must be able to handle increases in traffic or be expandable to handle increases when necessary.

DWDM technology provides us with an ability to expand our fiber network rapidly, meeting all the growing demands of the customer. DWDM, coupled with recent deployment of ATM switches, enables network simplification, network cost reductions, and new service offerings.

DWDM enables service providers to establish grow-as-you-go infrastructures. Providers can add current and new TDM systems to existing technologies, creating systems with endless expansion capacities. DWDM is considered to be a perfect fit for networks trying to meet bandwidth demands. DWDM systems are scalable. An example is a system of OC-48 interfacing with 16 channels per fiber line—this will enable the system to run efficiently as early as 2 years from now. It is projected that future DWDM terminals will carry up to 80 wavelengths of OC-48 or up to 40 wavelengths of OC-192, for a total of 200 or 400 Gbps, respectively. With enough capacity to transmit 90,000 volumes of an encyclopedia in 1 second, this will be a generous data storage boost.

Returning to our earlier car analogy, Figure 4-13 outlines how a service provider can integrate DWDM with its current technology, thereby increasing its system capacity.

Figure 4-13
DWDM integration

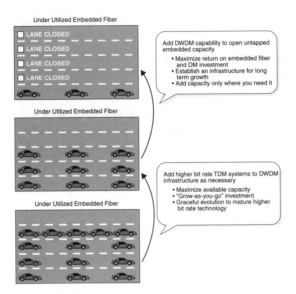

Ultra-wide-band optical-fiber amplifiers also may be used to boost light-wave signals carrying over 100 channels of light. By using amplifiers, a network can handle information in the terabit range. All the world's television channels or about half a million movies could be transmitted at the same time at this rate.

DWDM Costs

Three main needs drive every new technology. They are the desire for more, cheaper, and faster. Knowledge of DWDM is fascinating; academic arenas have studied this technology for years, but the real advances of this technology are promises for more data cheaper and faster. DWDM has been explained to have advantages with less amplification and easier ability to add to existing systems. These are advantages in cost reduction. Similar performance can be achieved using existing technologies, but costs would be high in comparison. DWDM costs can be broken down in comparison with SONET (TDM) and *Wave Division Multiplexing* (WDM)—long-haul versus short-haul data links.

DWDM is a subcategory of WDM. Both DWDM and WDM send data over multiple wavelengths of light. Data are sent in different channels of light. This is in contrast to a TDM system, which breaks different channels into time slots sent over one wavelength of light, when talking about fiberoptic transmissions. Long-haul links between high-speed carriers typically used TDM over optical fiber in the past. To initiate this, electrical regenerators must be installed every 40 to 100 km to boost the signal. The optical signal attenuates and requires a boost. Boosting the signal can be complicated. The TDM signal has strict timing requirements (it is a time division signal), so it must be reshaped, retimed, and then retransmitted. The equipment used to perform this function is expensive. It must be housed, powered, and maintained for proper operation. The propagation delay for the signal is increased because the optical signal is converted into an electrical signal and then reconverted into an optical signal. This means that less data can be transferred in the same amount of time. This equals more cost for the data provider. Over long distances, these factors increase the cost of using TDM dramatically.

WDM can be less costly in long-haul links. The optical signal attenuates and requires a boost, but this can be performed differently. WDM signals can use optical amplifiers that do not require costly electrical regeneration. Such a system uses EDFAs and can be spaced at a distance of up to 1,000 km. This is the second main advantage. The amplifiers do not need to demultiplex and process each signal. The optical amplifier simply amplifies the signals; it has no need to reshape, retime, or retransmit signals, and the process does not need to convert optical signals to electrical signals and back again. Combining simple amplification with increased distance between amplifiers results in dramatic savings. One optical amplifier on a 40-channel WDM system or a 150-channel DWDM system can replace 40 or 150 separate regenerators. This can be increased more because the distance between amplifiers is greater than between regenerators and because the signal is always transmitted as light so that amplifiers do not slow the link.

Another cost saving applies to long-haul carriers that already implement WDM over their links. The upgrade from an existing system to a system employing DWDM is relatively simple and cost-effective. One reason an upgrade such as this is simple has to do once again with the amplifiers used to boost the signal. The EDFAs already in place for WDM systems can amplify DWDM signals as well. Systems that employ TDM have an even greater cost associated with them when upgrades are necessary. Even more complicated electrical regenerators replace the old, and this equals an even greater cost.

A final cost advantage of DWDM over other systems is the ability to easily transmit several different protocols and speeds over the same link. A traditional TDM system, such as SONET, requires a total separate link or complicated and expensive conversion software to link routers and hubs with different protocols together. DWDM can send different protocols and different speed links over separate optical channels. One fiber can carry multiple protocols easily and cost-effectively.

Figure 4-14 is a graphic representation that compares DWDM systems with conventional TDM systems. Notice the OC-48 speeds on the transmit and receive sides can be replaced with higher or even different rates.

All the advantages of DWDM apply, for the most part, to long-haul links between major carriers. Most of the cost savings come from decreased needs to amplify or increased distances between amplifiers. In short-link applications, these savings cannot be realized. Both DWDM and TDM systems have no need to amplify under 40 km, and most LANs do not span any distance wider. In these cases, it is usually cheaper to lay more line or upgrade the existing

Figure 4-14
DWDM sample

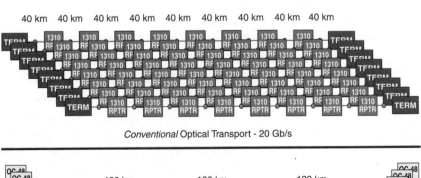
Conventional Optical Transport - 20 Gb/s

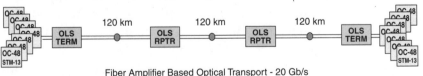
Fiber Amplifier Based Optical Transport - 20 Gb/s

TDM systems to increased capacities. DWDM routers necessary to implement the system are more expensive than the older SONET TDM systems. TDM systems that have been around longer are more readily available.

DWDM will be necessary in areas where extremely high traffic occurs over relatively large areas. An example of this is metropolitan New York and New Jersey. The need for an extremely high-speed network actually may exceed the capacity of current TDM systems.

Metropolitan Area Networks (MANs)

The Fibre Channel Industry Association recently introduced the proposed 10 Gb Fibre Channel standard for consideration. The 10 Gb Fibre Channel supports single- and multimode LAN and WAN devices over distances ranging from 15 m to 10 km. The standard supports bridging SANs over MANs through DWDM and SONET for disaster tolerance and data virtualization.

Work on the draft proposal to the American National Standards Institute began a year ago. The 10 Gb adapters, hubs, and storage arrays are anticipated early this year. Fibre Channel devices run at 1 Gbps, although 2 Gbps devices will start to ship by year-end. The 10 Gb draft requires backward compatibility with 1 and 2 Gbps devices. The 10 Gb devices will be able to use the same cable, connectors, and transceivers used in Ethernet and Infiniband. Infiniband is a new specification that provides a faster system bus between processors and peripherals. It is based on a switched matrix design.

Positioned as high-speed technology for networking applications in MANs; 10 Gb Ethernet provides very high bandwidths. It does so with simplicity and at relatively low costs compared with the only other MAN option, the traditional WAN infrastructure equipment. In MAN applications, 10 Gb Ethernet will enable service providers and network outsourcing providers to create very high-speed links at a low cost between colocated carrier-class switches and routers. This option for connection across MAN links will be at a data rate compatible with OC-192. This will enable Ethernet to extend to MAN networks and provide cost-effective solutions for larger geographic areas. For example, a nationwide Ethernet MAN carrier could link all its metropolitan areas in this way.

In 2000, Gigabit Ethernet was making inroads into the MAN space, and it paved the road for 10 Gb Ethernet. Startup MAN service providers have already adapted high-end Ethernet switches for their networks. Service providers make huge amounts of bandwidth cost-effective enough to run streaming video for customers with no problems. Not only do customers save on links to the MAN, but they also no longer need routers for connections outside their own LANs.

Gigabit Ethernet created such a good migration bridge to MANs, but why didn't this transition happen sooner? Fast Ethernet was too closely associated with copper and lacked the rudimentary *class of service* (CoS) that is available with new standards.

The fibreoptic Fast Ethernet standard has existed since 1995, but in the middle to late 1990s there wasn't much demand for fiber-based Fast Ethernet. Fiber-optic components were still expensive, and significant improvements in connector technology did not occur until later. Even Fast Ethernet would have been a relatively speedy connection out of a LAN, but there wasn't enough momentum to take it there.

Another impediment to 10 Gb Ethernet's adoption is the arrival of Gigabit Ethernet, roughly coinciding with other significant Ethernet advances. Layer 3 switch-based routing was just introduced, and layer 4 through layer 7 routing is not far behind. Both these added to the intelligence and speed of the LAN, outperforming traditional routers in this space. This widened the gap between LAN speeds and WAN speeds. In addition, fiberoptic component prices began to drop, making cost advantages of Gigabit Ethernet even more attractive. Application, management, storage, and other network service providers gained momentum as well. In the late 1990s, Fast Ethernet did not have enough bandwidth to offer outsourcing customers performance levels comparable with their existing LAN connections. Gigabit Ethernet arrived just in time to capitalize on another new trend: The installed base of dark fiber in major metropolitan areas was finally being lit up, and the investment environment for high-tech companies was high. This enabled companies to get funding to build MAN infrastructures from scratch.

Startup MAN service providers enjoyed an accounting edge over established competitors. As traditional carriers continued to invest in more expensive SONET/SDH and ATM equipment, they virtually locked themselves into those technologies owing to depreciation schedules. A traditional carrier that wanted to build an all-Ethernet MAN would have to buy all new equipment and take a huge loss on its existing MAN infrastructure. The Ethernet MAN market had to be created by startups. Such a financial risk for traditional carriers would not have been a consideration.

Many companies now are pursuing Ethernet in local loops, which will add to the explosion in bandwidth demand over carrier backbones, driving Ethernet further and further outside LANs. Enterprise networks that vastly outstrip WANs in terms of speed and residential markets are looking for similar speed boosts. With this kind of bandwidth demand focused on WANs, Ethernet MANs are being created swiftly to take advantage of the disparities.

Ethernet MANs have simplicity, ease of implementation, and relatively low cost on their side. Once established in major metropolitan areas, linking geographically close cities is only a few lines of fiber away. Regional area networks that run entirely on Ethernet could be created; the Northeast, northern California, and southern California would be logical starting points.

10 Gb Ethernet could be an effective link between cities and, with multiplexing, could outstrip OC-768 rates. Once these regional area networks are established, the momentum could move to the next rung, nationwide links, and from there to intercontinental links.

Wide Area Newtorks (WANs)

Ethernet's transition into the WAN space is ensured if its simplicity and low cost are as significant in the WAN as they have been in the LAN and MAN. The WAN historically has dominated traditional carriers with a voice background. Networks operate much closer to maximum load limits, cables are expected to run hundreds of kilometers without problems, and pricey equipment is the norm in this environment.

Back when the 155 Mbps OC-3 SONET system could easily outpace both 10 Mbps Ethernet and Fast Ethernet, it seemed that traditional WAN carriers had strong justifications for their thinking. Now data traffic far outweighs voice traffic, and this discrepancy is only expected to grow.

Data traffic does not require as much quality of service (QoS) as voice and video traffic. QoS is very dependent on how close traffic is running to the maximum load limit of the connection. In LAN thinking, it is time to get more bandwidth when you consistently see loads of 80 percent of the maximum available bandwidth across critical connections, usually in the backbone. In the LAN, extra bandwidth is relatively inexpensive and easy to add. In the WAN, bandwidth is expensive (1 Gbps of bandwidth in a WAN costs up to 10 times as much as its LAN equivalent), and installation is difficult. To provision a new network link, a network engineer must assess current capacity and reconfigure all cross-connects, adding and dropping multipliers along the route. This can take up to 60 days. Increasing SONET's scalability requires an upgrade for every device on the fiber route.

SONET's main weakness is that it has trouble handling data traffic. SONET systems were designed before the explosion of data services and equipment supporting traditional voice telecommunications payloads. Therefore, SONET traffic, whether voice, data, or video, must fit a specific bandwidth slot. In the case of data, the transmission may or may not use full bandwidth allocation. When payloads must be extended, SONET's poor signal concatenation schemes do not allow for full use of the payload, and the payload's bandwidth slot is left mostly empty. SONET is strictly a transport technology and unaware of payload traffic. It cannot address data transport requirements efficiently. Figure 4-15 depicts a SAN over MAN architecture.

This brings us to the WAN's major data transmission technology: ATM. Originally, ATM was considered the end-all solution for building a single topology across all networks. However, its complexity and high cost made it another victim of Ethernet in the LAN, and these same disadvantages could be its downfall in the WAN now that Ethernet has developed a competitive solution for that space.

The main arguments for ATM's continued success is its ability to provide true QoS and easy provisioning. In contrast with ATM's abil-

Configuring SANs: Dos and Don'ts

Figure 4-15
Extending SANs over MANs

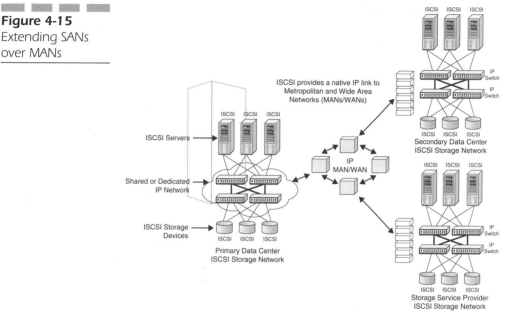

ity to guarantee that certain data will have certain levels of service, Ethernet, at its best, offers CoSs, basically as a best-effort service. ATM's flexible provisioning ensures that customers receive the bandwidths they pay for when they pay for them.

The advantages of ATM may be major pluses from a WAN standpoint; they are much less a factor in the LAN. With Ethernet's cheap and easy bandwidth, throwing extra bandwidth at them can solve QoS and provisioning problems. Throwing bandwidth at a problem is just not a traditional WAN way of solving a problem. And when the costs involved in adding extra bandwidth to a traditional WAN system are considered, this makes perfect sense. However, when your bandwidth prices for 1 Gbps of service are cut from one-fifth to one-tenth of customary levels, the "more bandwidth" approach looks more reasonable in the WAN.

The ability of 10 Gb Ethernet to compete with ATM and SONET opens up a market that has never been available to Ethernet. In 2000, combined ATM and SONET revenues were almost $21 billion. In 2004, the combined market will be $40 billion. However, the introduction of Ethernet could change these numbers drastically. From a revenue standpoint, the overall size of the WAN infrastructure market could fall as less-expensive Ethernet equipment floods the market. However, equipment shipment numbers could rise due to its low cost and ability to provide the extra bandwidth needed for QoS and easy provisioning. Refer to Figure 4-16 for a combined ATM and SONET market comparison.

The next 5 years will be interesting ones for the Ethernet world. 10 Gb Ethernet has the LAN locked up. The MAN space is already taking on overtones of the LAN battle part II. It seems that the LAN mindset is working just fine in the MAN space despite MAN's WAN roots. The WAN is where true magic may happen. 10 Gb Ethernet's use of SONET framing in the WAN physical layer asynchronous Ethernet interface sets the stage for 10 Gb Ethernet compatibility with all the traditional carriers' legacy equipment. This optional WAN physical layer interface includes a simple, inexpensive SONET framer; it operates at a data rate compatible with the payload rate of OC-192c/SDH VC-4-64c. There are significant invested interests in SONET and ATM, so don't expect these technologies to go down without a fight.

It is true that 10 Gb Ethernet has speed (100 Gb Ethernet could be a reality in as little as 4 years), price, simplicity, SONET compat-

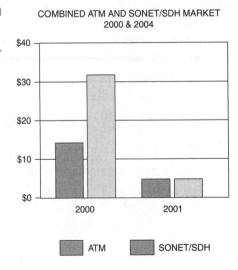

Figure 4-16
SONET versus ATM

ibility, and widespread industry support from LAN and WAN vendors. What it does not have is true QoS, provisioning, or the long-haul ranges of SONET.

The current advantages of 10 Gb Ethernet are almost the same as the strengths that enabled Ethernet to win over the LAN. These advantages are now allowing it to make inroads in the MAN. It remains to be seen if the WAN market's dynamics are different enough from those of the LAN to negate the effect of Ethernet's current strength. If the WAN market proves susceptible to Ethernet, which appears likely, 10 Gb Ethernet could help inject more reality into the dream of a united topology.

Asynchronous Transfer Mode (ATM)

ATM is a packet technology, meaning that all the information sent over an ATM network is broken down into packets. Unlike other packet technologies, ATM employs uniformly sized packets, called *cells,* each of which is 53 bytes long. Because the cells are all the same size, cell delay at ATM switches is more predictable and manageable. ATM was designed specifically to handle broadband applications efficiently and to let users have certain types of traffic priority treatment on the network. For example, voice traffic cannot tolerate much delay and can be marked as high priority with guaranteed bandwidth and minimal delays. Less sensitive traffic, e-mail, for instance, can be marked lower priority.

The first 5 bytes of an ATM cell, known as the *header,* contain an address telling where the cell wants to go in the network. The header specifies priority of the payload information carried in the remaining 48 bytes. ATM networks are linked together by a series of ATM switches that take in cells from various sources and switch them out again. There are two ways to establish connections between the switches. In a *permanent virtual circuit* (PVC), network managers preset paths from switch to switch ahead of time. PVCs were the initial way to establish virtual circuits between sites before the technology supporting *switched virtual circuits* (SVCs) was developed.

In a SVC, cells travel paths that are set up spontaneously and torn down again after designated periods. The advantage of this approach is that SVCs occupy switching capacity only when they are set up, increasing switch efficiency and payback on investments.

In either case, when a network device, such as a router, wants to send cells to an ATM network, it requests a virtual circuit. That request is passed from initial ATM switches to other ATM switches in the network, with each switch determining whether it can handle the request. If all switches along the path can accommodate the request, the virtual circuit is established. If not, the request must be repeated. This virtual circuit will support a certain QoS that has been preset in the ATM network switches based on user requirements. The most common ATM QoSs are known as *constant bit rate* (CBR), *variable bit rate* (VBR), *available bit rate* (ABR), and *unspecified bit rate* (UBR).

CBR guarantees low delay and the ability to handle a specified cell rate all the time. As long as the traffic falls within the cell rate, the network will handle it. It is intended for voice and high-quality video traffic. VBR is intended for bursty traffic. It allows traffic to burst up to specified rates for specified periods of time before discarding cells. VBR comes in two instances: real time and nonreal time. Real-time VBR limits delay that would hamper certain applications, such as some types of enterprise resource planning systems that require low network delay, plus some forms of video. Nonreal-time VBR is for bursty traffic that is unaffected by delay, such as more traditional LAN traffic or e-mail.

UBR is a best-effort QoS. When there is available bandwidth, the switch forwards UBR traffic. UBR is meant to support noncritical applications such as overnight file transfers. Initially, ATM was regarded as a way to send data faster than conventional means within a LAN. However, newer technologies such as Gigabit Ethernet offer the speed advantage without the drawbacks of ATM on the LAN.

One of these potential drawbacks is the amount of ATM traffic used for overhead. In addition to the 5 bytes of each 53-byte cell devoted to the header, in some cases 2 bytes or more are used for error checking—and some cells are used entirely for sending administrative data. All this can consume up to 20 percent of the throughput on a given circuit. This cell "tax" makes ATM an inefficient transport method for data-only networks where QoS is not an issue.

ATM has found a home on the backbone of large WANs and MANs. For example, carriers use ATM networks for transporting multiple types of traffic across long distances. Voice, Frame Relay, and other data traffic are converted into cells and sorted out at the other end of long-haul links.

In fact, large carriers are so heavily invested in ATM that they often mask its use under other names due to ATM's failure to achieve marketing buzz. Many more carrier services are ATM-based than users realize. For example, AT&T's IP-enabled Frame Relay is a service that attaches special labels representing groups of user IP addresses to create the illusion of an IP *virtual private network* (VPN)—it is really ATM between AT&T switches. SBC's Project Pronto is sold as a DSL connection to a nearby neighborhood gateway, but it is all ATM from there. Sprint's *Integrated On-Demand Network* (ION) is pure ATM from the customer premises device out.

Standards groups are working on ways to make ATM more attractive in LANs. *Multiprotocol over ATM* (MPoA) standards let non-ATM networks send traffic across an ATM network and tap into ATM QoS guarantees. ATM also has gained more flexibility in the WAN via recent standards. Some ATM carriers now allow users who cannot afford T-3 connections to choose multiples of T-1 bit rates via inverse multiplexing over ATM. And all the large carriers now allow users to keep Frame Relay equipment at some sites and ATM equipment at other sites in a single corporate network. This occurs via a standard called Frame Relay-to-ATM Interworking.

Few users will ever realize ATM's early promise of sending all types of traffic over a single network. This is so because ATM took too long to overcome early problems with expensive, hard-to-use equipment, and meanwhile, most efforts to create a single global network shifted to IP. ATM remains a key consideration for users needing to reduce disparate networks today while still enjoying reliable CoSs for different kinds of traffic.

Summary

SANs are growing rapidly because they solve problems. The problem is manageability of large and ever-increasing amounts of disk storage. When storage is attached directly to the computer through high-speed cable, the storage device can meet the needs of that computer very well. When computers are networked together, each computer with its own attached storage, total storage resource management is difficult or impossible to achieve.

The SAN approach offers many storage management advantages, including the ability to

- Share a pool of storage.
- Easily grow the size of the storage pool.
- Balance the load to each physical device.
- Eliminate downtime.
- Easily share backup devices.
- Back up without degrading performance for the network users.
- Manage storage resource easily and efficiently.
- Manage computing resource separately from storage resource.

The storage of digital data most of us take for granted as a kind of technological achievement that occurs independent of our recognition and operates solely as a behind-the-scenes process. Within the last 10 years, the Internet has grown from thousands of users to millions (53.5 million in 2001). SAN developments have increased the speed, reliability, and capacity of mass storage technologies, making it possible to save limitless quantities of information. The connections through which these systems exchange data do not always transmit mass quantities of information well.

Of the over 70 percent of mission-critical data stored on mainframes, the SAN community has all but ignored legacy systems. The clients, or end users, work from their workstations where stored data are backed up over the LAN. The LAN maintains UNIX, WINNT, Netware, or other legacy system servers types. Islands of SCSI disks process and provide first-level storage of data that are then backed up to a tape library system. LAN-free backups enable a SAN to share one of the most expensive components of the backup and recovery system, the tape or optical library and the drives within it.

The methodology shown in Figure 4-2 worked as long as a LAN could support network traffic and backups. Even with a state-of-the-art LAN, you will encounter individual backup clients that are too big to back up enterprise-wide across the LAN. Prior to building a SAN, you need to analyze your system requirements. Develop a SAN architecture that meets your system requirements and analysis. SAN architecture development will require good management in recognizing performance considerations in collecting data and implementing phase cycle development procedures successfully.

In Chapter 5, "Fibre Channel," we delve extensively into Fibre Channel and its products and cabling issues.

CHAPTER 5

Fibre Channel

Fibre Channel Overview

Fibre Channel is a set of clearly defined standards for communicating data. It is the best technology for applications that require high bandwidth. In addition to high bandwidth, features of Fibre Channel include

- Flexible topologies
- The ability of existing channel and networking protocols to run over the same physical interface and media
- Connectivity over several kilometers
- Support for multiple applications, including storage, network, video and clusters

Fibre Channel defines a high-speed protocol originally intended for superworkstations, large-array storage media, and high-performance desktop applications. Additionally, applications originally built using the *Small Computer System Interface* (SCSI) standard can use Fibre Channel. Fibre Channel supports multiple data rates of up to 4 Gbps in switched and shared architectures and in connection-oriented and connectionless modes. The physical media of Fibre Channel will support *Asynchronous Transfer Mode* (ATM).

Data communications may be categorized into two types, channels and networks. Channels are hardware intensive, whereas networks are software intensive. *Channels* are point-to-point links between communicating devices. Channels operate at hardware speeds, with minimal software overhead, and interconnect only a small number of devices over short distances. Examples of channel data communications include both SCSI and *High-Performance Parallel Interfaces* (HIPPIs).

Networks, on the other hand, provide low- to moderate-speed connections. Networks interconnect many devices, some of which may be physically distributed over long distances. Although networks have a higher software overhead, they are more flexible in supporting a variety of applications.

Channels are simple and provide higher performance and guaranteed delivery, whereas networks are more flexible at the cost of

Fibre Channel

lower throughput. Fibre Channel combines the desirable attributes of each. Fibre Channel networks can be configured in switched or redundant-loop topologies. This is defined by the *American National Standards Institute* (ANSI) and groups that support it, including the *International Committee for Information Technology Standards* (INCITS), formerly the National Committee for Information Technology Standards.

Concepts and Terminology

Before we begin a detailed discussion of Fibre Channel architecture, topology, classes, ports, and so on, let's take the time to familiarize ourselves with Fibre Channel concepts and terminology:

Acknowledgment frame (ACK)   An ACK is used for end-to-end flow control. An ACK is sent to verify receipt of one or more frames in class 1 and class 2 services.

Address identifier   A 3-byte value typically assigned by the fabric used to address an N_Port. Used in frames in the *source identifier* (S_ID) and *destination identifier* (D_ID) fields.

Arbitrated-loop physical address (AL_PA)   A 1-byte value used in the arbitrated-loop topology to identify L_Ports. This value will then also become the last byte of the address identifier for each public L_Port on the loop.

Arbitrated-loop timeout value (AL_TIME)   Twice the amount of time it would take for a transmission word to propagate around a worst-case loop, that is, a loop with 134 L_Ports and 10-km links between each L_Port with a six-transmission-word delay through each L_Port. This value is set at 15 ms.

Arbitrated loop   One of the three Fibre Channel topologies. Up to 126 NL_Ports and 1 FL_Port are configured in a unidirectional loop. Ports arbitrate for access to the loop based on their AL_PA. Ports with lower AL_PAs have higher priority than those with higher AL_PAs.

Arbitrate primitive signal (ARB)   This primitive signal applies only to the arbitrated-loop topology. It is transmitted as the fill word by an L_Port to indicate that the L_Port is arbitrating access to the loop.

Buffer-to-buffer credit value (BB_Credit)   Used for buffer-to-buffer flow control, this determines the number of frame buffers available in the port to which it is attached, that is, the maximum number of frames it may transmit without receiving an R_RDY.

Class 1 service   A method of communicating between N_Ports in which a dedicated connection is established between them. The ports are guaranteed the full bandwidth of the connection, and frames from other N_Ports may be blocked while the connection exists. In-order delivery of frames is guaranteed. Uses end-to-end flow control only.

Class 2 service   A method of communicating between N_Ports in which no connection is established. Frames are acknowledged by the receiver. Frames are routed through the fabric, and each frame may take a different route. In-order delivery of frames is not guaranteed. Uses both buffer-to-buffer flow and end-to-end flow control.

Class 3 service   A method of communicating between N_Ports similar to class 2 service, except that there is no acknowledgment of received frames. Frames are routed through the fabric as in class 2 service, and in-order delivery is not guaranteed. Uses only buffer-to-buffer flow control.

Close primitive signal (CLS)   This primitive signal applies only to the arbitrated-loop topology. A primitive signal is sent by an L_Port that is currently communicating on the loop (that is, it has won access to the loop or was opened by another L_Port that had won access to the loop) to close communication with the other L_Port.

D_ID   A 3-byte field in the frame header used to indicate the address identifier of the N_Port to which the frame is to be delivered.

Fibre Channel

Disparity   The difference between the number of 1s and 0s in a transmission character. A transmission character with more 1s than 0s is said to have *positive-running disparity*. A transmission character with more 0s than 1s is said to have *negative-running disparity*. A transmission character with an equal number of 1s and 0s is said to have *neutral disparity*.

Error detect timeout value (E_D_TOV)   A timer used to represent the longest possible time for a frame to make a roundtrip through the fabric. This value is negotiated at N_Port login and typically will be on the order of a few seconds. E_D_TOV is used to decide when some particular error-recovery action must be taken.

End-to-end credit value (EE_Credit)   Used for end-to-end flow control to determine the maximum number of frames that may remain unacknowledged.

End-of-frame delimiter (EOF)   This ordered set is always the last transmission word of a frame. It is used to indicate that a frame has ended and indicates whether the frame is valid or invalid.

ESCON   Mainframe channel devised by IBM to support data rates of 200 Mbps over fiber. Depending on configuration, ESCON can reach distances from about 3 to 10 km over fiber-quality product features.

Exchange   The highest-level Fibre Channel mechanism used for communication between N_Ports. Exchanges are composed of one or more related sequences. Exchanges may be bidirectional or unidirectional.

Fabric   One of the three Fibre Channel topologies. In the fabric topology, N_Ports are connected to F_Ports on a switch. Depending on vendor support, fabric switches may be interconnected to support up to 16 million or more N_Ports on a single network.

Fabric port busy frame (F_BSY)   This frame is issued by the fabric to indicate that a particular message cannot be delivered because the fabric or the destination N_Port is too busy.

FICON   Mainframe channel devised by IBM that supports data rates of 1 Gbps at distances of about 10 to 20 km over fiber.

Fibre Channel arbitrated loop (FC-AL)   Refers to the ANSI FC-AL document that specifies operation of the arbitrated-loop topology.

Fibre Channel physical and signaling interface (FC-PH)   Refers to the ANSI FC-PH document that specifies the FC-0, FC-1, and FC-2 layers of the Fibre Channel protocol.

Fibre Channel layer 0 (FC-0)   Specifies the physical signaling used in Fibre Channel, as well as cable plants, media types, and transmission speeds.

Fibre Channel layer 1 (FC-1)   Specifies the IBM-patented 8B/10B data encoding used in Fibre Channel.

Fibre Channel layer 2 (FC-2)   Specifies the frame format, sequence/exchange management, and ordered set usage in Fibre Channel.

Fibre Channel layer 3 (FC-3)   Specifies services provided for multiple N_Ports in a single node.

Fibre Channel layer 4 (FC-4)   Specifies mapping of upper-level protocols such as SCSI and *Internet Protocol* (IP) onto the Fibre Channel Protocol.

Fill word   The primitive signal used by L_Ports to be transmitted in between frames. This may be idle or ARBx depending on which, if any, L_Ports are arbitrating for loop access and will not necessarily be the same for all L_Ports on the loop at any given time.

Fabric-loop port (FL_Port)   An F_Port that is capable of supporting an attached arbitrated loop. An FL_Port on a loop will have the AL_PA hex'00', giving the fabric highest-priority access to the loop. An FL_Port is the gateway to the fabric for NL_Ports on a loop.

F/NL_Port   An NL_Port that is capable of providing certain fabric services to other NL_Ports on a loop in the absence of a fabric. This NL_Port will respond to requests to open communication with AL_PA hex'00' even though it actually may have another value for its AL_PA.

Fabric port (F_Port)   A port on a fabric switch to which N_Ports may be connected directly. An F_Port uses the address identifier hex'FFFFFE'.

Frame   The basic unit of communication between two N_Ports. Frames are composed of a starting delimiter (SOF), a header, the payload, the *cyclic redundancy check* (CRC), and an ending delimiter (EOF). The SOF and EOF contain the special character and are used to indicate where the frame begins and ends. The 24-byte header contains information about the frame, including the S_ID, D_ID, routing information, the type of data contained in the payload, and sequence/exchange management information. The payload contains the actual data to be transmitted and may be 0 to 2,112 bytes in length. The CRC is a 4-byte field used for detecting bit errors in the received frame.

Fabric port reject frame (F_RJT)   This frame is issued by the fabric to indicate that delivery of a particular frame is being denied. Some reasons for issuing an F_RJT include class not supported, invalid header field(s), and N_Port unavailable.

Fibre Channel over IP (FCIP)   A proposal advanced by IETF's IP Storage Working Group for mapping Fibre Channel traffic directly onto IP in order to facilitate *wide area network* (WAN) connectivity for *storage area networks* (SANs).

Fiber-optic cable connector   The SC connector is the standard connector for Fibre Channel fiber-optic cable. It is a push-pull connector favored over the ST connector. If the cable is pulled, the tip of the cable in the connector does not move out, resulting in loss of signal quality.

Idle   An ordered set transmitted continuously over a link when no data are being transmitted. Idle is transmitted to maintain an active link over a fiber and lets the receiver and transmitter maintain bit, byte, and word synchronization.

Intermix   A service in which class 2 and class 3 frames may be delivered to an N_Port that has a class 1 dedicated connection open. The class 2 and 3 frames are delivered during times when no class 1 frames are being delivered on the connection.

Link service   Link services are facilities used between an N_Port and a fabric or between two N_Ports for such purposes as login, sequence and exchange management, and maintaining connections.

Loop initialization primitive sequence (LIP)   This primitive sequence applies only to the arbitrated-loop topology. It is transmitted by an L_Port to (re)initialize the loop.

Loop initialization fabric assigned frame (LIFA)   This is the first frame transmitted in the loop initialization process after a temporary loop master has been selected. L_Ports that have been assigned their AL_PA by the fabric will select their AL_PAs in this frame as it makes its way around the loop.

Loop initialization hard assigned frame (LIHA)   This is the third frame transmitted in the loop initialization process after a temporary loop master has been selected. L_Ports that have been programmed to select a particular AL_PA (if available) by the manufacturer will select their AL_PAs in this frame as it makes its way around the loop.

Loop initialization loop position frame (LILP)   This is the second frame transmitted in the loop initialization process after all L_Ports have selected an AL_PA (after LISA has been around the loop). This frame is transmitted around the loop so that all L_Ports may know the relative position of all other L_Ports around the loop. Support for this frame by an L_Port is optional.

Loop initialization previously assigned frame (LIPA)   This is the second frame transmitted in the loop initialization process after a temporary loop master has been selected. L_Ports that had an AL_PA prior to the loop initialization will select their AL_PAs in this frame as it makes its way around the loop.

Loop initialization report position frame (LIRP) This is the first frame transmitted in the loop initialization process after all L_Ports have selected an AL_PA (after LISA has been around the loop). This frame is transmitted around the loop so that all L_Ports report their relative physical position on the loop. Support for this frame by an L_Port is optional.

Loop initialization soft assigned frame (LISA) This is the fourth frame transmitted in the loop initialization process after a temporary loop master has been selected. L_Ports that did not select an AL_PA in any of the previous loop initialization frames (LIFA, LIPA, or LIHA) will select their AL_PAs in this frame as it makes its way around the loop.

Loop initialization select master frame (LISM) This frame applies only to the arbitrated-loop topology. It is the first frame transmitted in the initialization process in which L_Ports select an AL_PA. It is used to select a temporary loop master or the L_Port that subsequently will initiate transmission of the remaining initialization frames (LIFA, LIPA, LIHA, LISA, LIRP, and LILP).

Loop port bypass primitive sequence (LPB) This primitive sequence applies only to the arbitrated-loop topology. It is transmitted by an L_Port to bypass the L_Port to which it is directed. For example, if port A suspects that port B is malfunctioning, port A can send an LPB to port B so that port B will only retransmit everything it receives and will not be active on the loop.

Loop port enable primitive sequence (LPE) This primitive sequence applies only to the arbitrated-loop topology. It is transmitted by an L_Port to enable an L_Port that has been bypassed with the LPB primitive sequence.

Loop port (L_Port) Generic term for an NL_Port or FL_Port, that is, any Fibre Channel port that supports the arbitrated-loop topology.

Loop port state machine (LPSM) This is a state machine maintained by an L_Port to track its behavior through different phases of loop operation, that is, how it behaves when it is arbitrating for loop access, how it behaves when it has control of the loop, and so on.

Link reset primitive sequence (LR)   This primitive sequence is used during link initialization between two N_Ports in the point-to-point topology or an N_Port and an F_Port in the fabric topology. The expected response to a port sending LR is the LRR primitive sequence.

Link reset response primitive sequence (LRR)   This primitive sequence is used during link initialization between two N_Ports in the point-to-point topology or an N_Port and an F_Port in the fabric topology. It is sent in response to the LR primitive sequence. The expected response to a port sending LRR is idle.

Mark primitive signal (MRK)   This primitive signal applies only to the arbitrated-loop topology and is transmitted by an L-PORT for synchronization purposes. It is vendor specific.

Node-loop port (NL_Port)   An N_Port that can operate on the arbitrated-loop topology.

Nonparticipating mode   An L_Port will enter the nonparticipating mode if there are more than 127 devices on a loop, and it thus cannot acquire an AL_PA. An L_Port also may enter the nonparticipating mode voluntarily if it is still physically connected to the loop but wishes not to participate. An L_Port in the nonparticipating mode is not capable of generating transmission words on the loop and may only retransmit words received on its inbound fiber.

Not operational primitive sequence (NOS)   This primitive sequence is used during link initialization between two N_Ports in the point-to-point topology or an N_Port and an F_Port in the fabric topology. It is sent to indicate that that transmitting port has detected a link failure or is offline. The expected response to a port sending NOS is the OLS primitive sequence.

Node port (N_Port)   A port on a computer or disk drive through which the device does its Fibre Channel communication.

N_Port name   An 8-byte manufacturer-assigned value that uniquely identifies the N_Port throughout the world.

Offline primitive sequence (OLS)   This primitive sequence is used during link initialization between two N_Ports in the point-to-point topology or an N_Port and an F_Port in the fabric topology. It is sent to indicate that the transmitting port is attempting to initialize the link, has recognized the NOS primitive sequence, or is going offline. The expected response to a port sending OLS is the LR primitive sequence.

Open primitive signal (OPN)   This primitive signal applies only to the arbitrated-loop topology. The OPN primitive signal is sent by an L_Port that has won the arbitration process to open communication with with one or more other ports on the loop.

Ordered set   A 4-byte transmission word that has the special character as its first transmission character. An ordered set may be a frame delimiter, a primitive signal, or a primitive sequence. Ordered sets are used to distinguish Fibre Channel control information from data.

Originator   The N_Port that originated an exchange.

Originator exchange identifier (OX_ID)   A 2-byte field in the frame header used by the originator of an exchange to identify frames as being part of a particular exchange.

Participating mode   The normal operating mode for an L_Port on a loop. An L_Port in this mode has acquired an AL_PA and is capable of communicating on the loop.

Primitive sequence   An ordered set transmitted repeatedly and used to establish and maintain a link. LR, LRR, NOS, and OLS are primitive sequences used to establish an active link in a connection between two N_Ports or an N_Port and an F_Port. LIP, LPB, and LPE are primitive sequences used in the arbitrated-loop topology for initializing the loop and enabling or disabling an L_Port.

Primitive signal   An ordered set used to indicate an event. Idle and R_RDY are used in all three topologies. ARB, OPN, CLS, and MRK are used only in the arbitrated-loop topology.

Private loop   An arbitrated loop that stands on its own; that is, it is not connected to a fabric.

Private NL_Port   An NL_Port that only communicates with other ports on the loop, not with the fabric. Note that a private NL_Port may exist on either a private or a public loop.

Public loop   An arbitrated loop that is connected to a fabric.

Public NL_Port   An NL_Port that may communicate with other ports on a loop as well as through an FL_Port to other N_Ports connected to the fabric.

Responder   The N_Port with which an exchange originator wishes to communicate.

Responder exchange identifier (RX_ID)   A 2-byte field in the frame header used by the responder of an exchange to identify frames as being part of a particular exchange.

Sequence   A group of related frames transmitted unidirectionally from one N_Port to another.

Sequence identifier (SEQ_ID)   A 1-byte field in the frame header used to identify which sequence of an exchange a particular frame belongs to.

Sequence initiator   The N_Port that began a new sequence and transmits frames to another N_Port.

Sequence recipient   The N_Port to which a particular sequence of data frames is directed.

S_ID   A 3-byte field in the frame header used to indicate the address identifier of the N_Port from which the frame was sent.

Start-of-frame delimiter (SOF)   This ordered set is always the first transmission word of a frame. It is used to indicate that a frame will immediately follow and indicates which class of service the frame will use.

Special character   A special 10-bit transmission character that does not have a corresponding 8-bit value but is still considered valid. The special character is used to indicate that a particular transmission word is an ordered set. The special character is the only transmission character to have five 1s or 0s in a row. The special character is also referred to as K28.5 when using K/D format.

Transmission character   A (valid or invalid) 10-bit character transmitted serially over the fiber. Valid transmission characters are determined by the 8B/10B encoding specification.

Transmission word   A string of four consecutive transmission characters.

Upper layer protocol (ULP)   The protocol that runs on top of Fibre Channel through the FC-4 layer. Typical ULPs running over Fibre Channel are SCSI, IP, HIPPI, and *Intelligent Peripheral Interface* (IPI).

Additional Fibre Channel terminology appears at the end of this chapter, along with illustrations.

Fibre Channel Architecture

Data in a Fibre Channel environment are communicated through five layers of architecture: FC-0, FC-1, FC-2, FC-3, and FC-4.

FC-0 and FC-1 can be thought of as defining the physical layer of the *Open Systems Interconnection* (OSI) model. FC-2 is similar to what other protocols define as a *media access control* (MAC) layer, which is typically the lower half of the data link layer. Fibre Channel, however, does not define the concept of a MAC. FC-3 is not really a layer at all. It is still a largely undefined set of services for devices having more than one port. An example is striping, where data are transmitted out of all ports at the same time in order to increase bandwidth. FC-4 defines how other well-known higher-layer protocols are mapped onto and transmitted over Fibre Channel.

The three bottom layers of the Fibre Channel stack (FC-0 through FC-2) form what is known as the *Fibre Channel physical standard* (FC-PH). This defines all the physical transmission characteristics of Fibre Channel. The remaining layers (FC-3 and FC-4) handle interfaces with other network protocols and applications.

Unlike *local area network* (LAN) technologies such as Ethernet and Token Ring, Fibre Channel keeps the various functional layers of the stack physically separate. This enables vendors to build products from discrete functional components, such as chip sets and bus interfaces, that handle specific portions of the protocol stack. It also enables them to implement some stack functions in hardware and others in software or firmware.

FC-0 Layer: Physical Layer

FC-0 defines the basic physical link, including the fiber, connectors, and optical/electrical parameters for a variety of data rates. Fibre Channel offers a basic rate of 133 MBd, the most commonly used speed of 266 MBd, and 531 MBd and 1.062 GBd. It is important to note that these signaling rates include the overhead involved in establishing and maintaining connections. The actual data throughput is somewhat lower: 100 Mbps for 133 MBd, 200 Mbps for 266 MBd, 400 Mbps for 531 MBd, and 800 Mbps for 1.062 GBd.

Data rates are expected to go well beyond today's 1.062-GBd ceiling. In most cases, including switched environments, users will be able to upgrade simply by swapping in an adapter card. Specifications for 2.134 and 4.268 GBd are already in place.

FC-1 Layer: Transmission Encode/Decode Layer

FC-1 defines the transmission protocol, including serial encoding and decoding rules, special characters, timing recovery, and error control. The information transmitted over a fiber is encoded 8 bits at a time into a 10-bit transmission character (8B/10B transmission coding scheme licensed from IBM). The two extra bits are used for error detection and correction, known as *disparity control*. The 8B/10B scheme supplies sufficient error detection and correction to permit use of low-cost transceivers, as well as timing recovery methods to reduce the risk of radiofrequency interference and ensure balanced, synchronized transmissions.

FC-2 Layer: Framing and Signaling Layer

The FC-2 layer performs the basic signaling and framing functions and defines the transport mechanism for data from the upper layers of the stack.

FC-2 frames and sequences data from the upper layers for transmission via the FC-0 layer. It also accepts transmissions from the FC-0 layer and reframes and resequences them for use by the upper layers. Fibre Channel frame sizes can vary and must be negotiated by the transmitter-receiver pair for each connection. Frame sizes typically range from 36 bytes to 2 kilobytes but may be larger in some cases. Fibre Channel is making the biggest impact in the storage arena particularly using SCSI as the ULP. Compared with traditional SCSI, the benefits of mapping SCSI command set onto Fibre Channel include

- Faster speed
- More device connections
- Larger distances allowable between devices

Fibre Channel using arbitrated-loop topology is simply a replacement for SCSI. Many companies are moving toward SCSI adapter cards for several platforms and operating systems in addition to disk drives and storage devices with Fibre Channel interfaces. SCSI companies are selling Fibre Channel devices that run IP. Although ULP is an independent topology, IP is found more commonly in switched fabric environments. At the present time, SCSI and IP are pretty much the only two ULPs used commercially on Fibre Channel.

FC-0 layer features include

- Signaling
- Media specifications
- Receiver-transmitter specifications

FC-2 signaling defines the connection between at least two Fibre Channel node ports (or N_Ports). One of these acts as originator of outbound traffic, and one other acts as a responder receiving inbound traffic and sending traffic back. Connections between N_Ports are in full duplex. The FC-2 layer provides traffic management functions, including flow control, link management, and buffer memory management, in addition to error detection and correction.

The FC-2 layer supplies additional flow control and error correction mechanisms. For flow control, it employs sliding-window schemes similar to that of *Transmission Control Protocol* (TCP)/ *Internet Protocol* (IP). Flow control functionality varies somewhat according to class of service defined. Error detection is performed using a 32-bit CRC. A *link control facility* (LCF) manages the Fibre Channel connection and maintains the information needed to recover links in the event of failure. FC-2 features include

- Frame format
- Sequence management
- Exchange management
- Flow control
- Classes of service
- Login/logout
- Segmentation and reassembly

FC-3 Layer: Advanced Features

The FC-3 layer defines special service features, including

- *Striping*, which is used to multiply bandwidth using multiple N_Ports in parallel to transmit a single information unit across multiple links.
- *Hunt groups*, which provide the ability for more than one port to respond to the same alias address. This improves efficiency by decreasing the chance of reaching a busy N_Port.
- *Multicast*, which delivers a single transmission to multiple destination ports. This includes sending to all N_Ports on a fabric (broadcast) or to only a subset of the N_Ports on a fabric.

Fibre Channel

FC-4 Layer: Protocol Mapping

The FC-4 layer specifies the mapping rules for several legacy ULPs. FC-4 enables Fibre Channel to carry data from other networking protocols and applications. Fibre Channel can concurrently transport both network and channel information over the same physical interface. The following network and channel protocols are specified as FC-4:

- ULP mapping
- SCSI
- IP
- HIPPI
- *ATM Adaption Layer 5* (ATM-AAL5)
- *Intelligent Peripheral Interface-3* (IPI-3) (disk and tape)
- *Single Byte Command Code Sets* (SBCCS)
- ULPs
- IPI
- HIPPI Framing Protocol
- *Link Encapsulation* (FC-LE)
- SBCCS mapping
- IEEE 802.2

Ports

There are three basic types of ports in Fibre Channel: the N_Port, F_Port, and E_Port. As you add arbitrated-loop capabilities, these basic ports become the NL_Port, FL_Port, and G_Port:

> *N_Port*   An N_Port is a node port, or a port on a disk or computer. A port that is only an N_Port can only communicate with another N_Port on another node or to a switch.

F_Port   An F_Port is a fabric port, which is found only on a switch. A port that is only an F_Port can only connect to another N_Port via a point-to-point connection.

L_Port   The term L_Port is not really used in Fibre Channel, but the *L* in the name implies that it is a port that can participate in an arbitrated loop. Typically, this *L* is added to the end of an N_Port or F_Port to create an NL_Port or an FL_Port.

NL_Port   An N-Port with arbitrated-loop capabilities, that is, a port on a node that can connect to another node or a switch (see the N_Port definition), or it can participate in an arbitrated loop (see the L_Port definition).

FL_Port   An F-Port with arbitrated-loop capabilities, that is, a port on a switch that can connect to a node (see the F_Port definition) or an arbitrated loop (see the L_Port definition).

E_Port   An E_Port is an expansion port on a switch that is used to connect to other switches via their E_Ports to form a large fabric.

G_Port   A G_Port is a generic port on a switch that can act as an E_Port, FL_Port, or F_Port depending on what connects to it.

As the transmission interface between servers and clustered storage, Fibre Channel switches are more expensive than Ethernet networking. The price of Ethernet and related products has decreased as the mass marketing of *personal computers* (PCs) equipped with Ethernet ports and networking products has resulted in prices at which they could enter the home.

Fibre Channel Service Classes

FC-2 defines classes of service as unique features of Fibre Channel that enable it to meet a variety of communications needs. The user selects a service class based on characteristics of his or her application, such as packet length and transmission duration. FC-2 allocates services by the fabric login protocol. Discussion of the service classes follows.

Class 1

In class 1, a dedicated connection is established between two N_Ports. Once established, the two N_Ports may communicate using the full bandwidth of the connection; no other network traffic affects this communication. Because of this, frames are guaranteed to arrive in the order in which they were transmitted. In addition, the media speeds must be the same for all links that make up the dedicated connection. Because of the nature of the dedicated connection, there is no need for buffer-to-buffer flow control; the fabric does not need to buffer the frames as they are routed. Thus, only end-to-end flow control is used in class 1. Class 1 would be used when the data must be continuous and time critical, such as voice or video.

Hard-wired or circuit-switched connections are dedicated, uninterruptible links such as telephone connections. Features include

- A connection that is retained and guaranteed by the fabric
- Guaranteed maximum bandwidth between the two N_Ports
- Frame delivery to the destination port in the same order as transmitted
- Sustained, high-throughput transactions
- Time-critical, nonbursty, dedicated connections, such as those between two supercomputers

Class 2

Class 2 is referred to as *multiplex* because it is a connectionless class of service with notification of delivery and nondelivery of frames. Since no dedicated connection needs to be established, a port can transmit frames to and receive frames from more than one N_Port. As a result, the N_Ports share the bandwidth of the links with other network traffic. Frames are not guaranteed to arrive in the order in which they were transmitted, except in the point-to-point or loop topologies. In addition, the media speeds may vary for different links that make up the path. Both buffer-to-buffer and end-to-end flow controls are used in class 2. Class 2 is more like typical LAN traffic, such as IP or *File Transfer Protocol* (FTP), where the order and timeliness of delivery are not so important.

Connectionless, frame-switched transmission guarantees delivery and confirms receipt of traffic. Features include

- Functions that are performed on the data frame rather than on a connection
- No dedicated connection between N_Ports
- Each frame transmitted to its destination over any available route
- No guarantee of the order of delivery of frames
- Retransmittal of a frame when congestion occurs until the frame reaches its destination successfully

In addition, when class 2 signaling blocks a frame, the transmitting N_Port receives a busy frame. This enables retransmission to be performed immediately rather than forcing the station to wait, as is the case with TCP/IP and other protocols.

Class 3

Class 3 is very similar to class 2. The only exception is that it only uses buffer-to-buffer flow control. It is referred to as a *datagram service*. Class 3 would be used when order and timeliness are not so important and when the ULP itself handles lost frames efficiently. Class 3 is the choice for SCSI and includes the following features:

- One-to-many connectionless, frame-switched service
- Class 2-like service but no delivery guarantee or confirmation mechanism

Fibre Channel

- Greater speed than class 2 because there is no wait for confirmation
- No retransmittal if a transmission does not arrive at its destination

Class 3 service is used most often for real-time broadcasts that cannot wait for acknowledgment but are not sufficiently time-critical to warrant class 1 service. It is also used for applications that can tolerate lost packets.

Class 4

Class 4 provides fractional bandwidth allocation of the resources of a path through a fabric that connects two N_Ports. Class 4 can be used only with the pure fabric topology. One N_Port will set up a *virtual circuit* (VC) by sending a request to the fabric indicating the remote N_Port as well as quality-of-service parameters. The resulting class 4 circuit will consist of two unidirectional VCs between the two N_Ports. The VCs need not be the same speed.

Like a class 1 dedicated connection, class 4 circuits will guarantee that frames arrive in the order in which they were transmitted and will provide acknowledgement of delivered frames (class 4 end-to-end credit). The main difference is that an N_Port may have more than one class 4 circuit, possibly with more than one other N_Port at the same time. In a class 1 connection, all resources are dedicated to the two N_Ports. In class 4, the resources are divided up into potentially many circuits. The fabric regulates traffic and manages buffer-to-buffer flow control for each VC separately using the FC_RDY primitive signal. Intermixing of class 2 and 3 frames is mandatory for devices supporting class 4.

Class 5

The idea for class 5 involved isochronous, just-in-time service. However, it is still undefined.

Class 6

Class 6 provides support for multicast service through a fabric. Basically, a device wanting to transmit frames to more than one N_Port at a time sets up a class 1 dedicated connection with the multicast server within the fabric at the well-known address of hex'FFFFF5'. The multicast server sets up individual dedicated connections between the original N_Port and the entire destination N_Ports. The multicast server is responsible for replicating and forwarding the frame to all other N_Ports in the multicast group. N_Ports become members of a multicast group by registering with the alias server at the well-known address of hex'FFFFF8'. Class 6 is very similar to class 1; class 6 SOF delimiters are the same as those used in class 1. In addition, end-to-end flow control is used between the N_Ports and the multicast server.

Intermix

Intermix is an option of class 1 whereby class 2 and 3 frames may be transmitted at times when class 1 frames are not being transmitted. The class 2 and 3 frames may or may not be destined to the same N_Port as the class 1 frames. Both N_Ports and the fabric must support intermix for it to be used.

Fibre Channel Topologies

Fibre Channel defines three topologies: point to point, arbitrated loop, and fabric. Their descriptions follow.

Point to Point

Point-to-point topology is the simplest of the three. It consists of two (and only two) Fibre Channel devices connected directly together. The transmit fiber of one device goes to the receive fiber of the other device, and vice versa. There is no sharing of media that enables devices to enjoy the total bandwidth of the link. A simple link initialization is required of the two devices before communication begins. Point-to-point topology features include

- Simplicity
- Bidirectional links that interconnect the N_Ports of a pair of nodes
- Nonblocking
- Underutilized bandwidth of the communications link

Arbitrated Loop

Arbitrated loop has become the most dominant Fibre Channel topology and is the most complex. It is a cost-effective way of connecting up to 127 ports in a single network without the need of a fabric switch. Unlike the other topologies, the media are shared among the devices, limiting each device's access. Not all devices are required to operate on an arbitrated loop; added functionality is optional. For a loop to operate, all devices must be loop devices.

Arbitrated loop is not a token-passing scheme. When a device is ready to transmit data, it first must arbitrate and gain control of the loop. It does this by transmitting the *arbitrate* (ARBx) primitive signal, where x is the *arbitrated-loop physical address* (AL_PA) of the device. Once a device receives its own ARBx primitive signal, it has gained control of the loop and can communicate with other devices by transmitting an *open* (OPN) primitive signal to a destination device. When this occurs, there exists point-to-point communication between the two devices. All other devices simply repeat the data.

If more than one device on a loop is arbitrating at the same time, the x values of the ARBx primitive signals are compared. When an arbitrating device receives another device's ARBx, the ARBx with the numerically lower AL_PA is forwarded, and the ARBx with the numerically higher AL_PA is blocked. The device with the lower AL_PA will gain control of the loop first. That device relinquishes control of the loop, and the other device has a chance.

Unlike token-passing schemes, there is no limit on how long a device may retain control of the loop. This demonstrates the channel aspect of Fibre Channel. There is an access fairness algorithm that prohibits a device from arbitrating until all other devices have had a chance to arbitrate. The catch is that the access fairness algorithm is optional. Arbitrated loop

- Is basically the simplest form of a fabric topology
- Has shared bandwidth, distributed topology
- Interconnects NL_Ports and FL_Ports at the nodes/fabric using unidirectional links
- Allows only one active L_Port-to-L_Port connection, so blocking is possible
- Has a fairness algorithm that ensures that no L_Port is blocked from accessing the loop
- Terminates communication between all L_Ports if any link in the loop fails
- Does not need a switched fabric and so is less expensive than cross-point
- Provides a bidirectional connection between a node (N_Port) and the fabric (F_Port)
- Can be configured to be nonblocking by providing multiple paths between any two F_Ports
- Uses a D_ID, embedded in the frame header, to route the frame through to the desired destination N_Port
- Provides the highest performance and connectivity of the three topologies
- Provides efficient sharing of available bandwidth
- Guarantees connectivity without congestion
- Allows the addition of stations to a fabric without reducing the point-to-point channel bandwidth

Loop Initialization Before the loop is usable, it must be initialized so that each port obtains an AL_PA, a dynamically assigned value by which the ports communicate. It maps to the lowest byte of the native address identifiers (_ID and SI_ID). Although the AL_PA is 1 byte long, only 127 values are valid (neutral running disparity).

If more than 127 devices are present on the loop, some will not be able to select an AL_PA and will be forced into nonparticipating mode. The basics of loop initialization are as follows:

1. The LIP primitive sequence begins the process. An L_Port transmits LIP after it powers on or when it detects loop failure (loss of synchronization at its receiver). The LIP will propagate around the loop, triggering all other L_Ports to transmit LIP as well. At this point, the loop is not usable.

2. The next major step is to select a loop master that will control the process of AL_PA selection. This is done by the L_Ports constantly transmitting LISM frames. The process is designed so that if a fabric is present, it will become loop master (by receiving back its own LISM frames); otherwise, the port with the numerically lowest port name will win. All other L_Ports propagate the higher-priority LISM frames.

3. The third step is to enable the L_Ports to select an AL_PA. The concept of an AL_PA bitmap is used, where each L_Port selects (and sets) a single bit in the bitmap of a frame originated by the loop master and repeats the frame back on the loop. There are 127 available bits, corresponding to the 127 valid AL_PAs. This process is done using four frames, breaking the selection down according to priority.

If an L_Port has a valid AL_PA before the loop began initialization, it will attempt to reclaim this previously acquired value by looking for that bit to be available in the LIPA frame. If it is, it will set the bit and repeat the frame. If it is not available (already been claimed), the L_Port will wait for the LISA frame to come around and claim one there.

Once the LISA frame has come back to the loop master, all L_Ports (hopefully) have selected an AL_PA. The loop master may send two additional frames, but only if all L_Ports on the loop support them. The first is the *loop initialization report position* (LIRP). As the frame traverses the loop, each port adds its AL_PA to the end of a list. When this is done, the relative positions of all L_Ports are known. Finally, the *loop initialization loop position* (LILP) frame is transmitted, which simply enables all L_Ports to look at the finished list. Whether or not LIRP and LILP are used, the loop master transmits the CLS primitive signal to let each port know that the process has finished. At this point, the loop has finished initializing and is ready to be used.

Fabric

The fabric topology is used to connect many devices in a cross-point-switched configuration. The benefit of this topology is that many devices can communicate at the same time—the media are not shared. It also requires the purchase of a switch. When the N_Ports log into the fabric, the fabric will assign native address identifiers (S_ID). Functions of the fabric include a multicast server, broadcast server, alias server, quality-of-service facilitator, and directory server. Some fabrics have FL_Ports that enable arbitrated loops to be connected to the fabric.

More Concepts and Terminology

Additional Fibre Channel terms and illustrations include the following.

Copper Cables

Four kinds of copper cables are defined in the Fibre Channel standard. The most popular implementations are twin-axial using DB-9 or HSSD connectors.

Disk Enclosures

Fibre Channel disk enclosures use a back plane with built-in Fibre Channel loops. At each disk location, in the back-plane loop, is a port bypass circuit that permits hot swapping of disks. If a disk is not present, the circuit automatically closes the loop. When a disk is inserted, the loop is opened to accommodate the disk.

Drivers

If software drivers for the *host bus adapter* (HBA) vendor are not resident in your server or workstation, they are installed into the operating system using standard procedures for the operating system. Fibre Channel drivers support multiple protocols, typically SCSI and IP. Most popular operating systems are supported, including Windows NT, AIX, Solaris, IRIX, and HPUX.

Extenders

Extenders are used to provide longer cable distances. Most optical interfaces are multimode cable. Extenders convert the multimode interface to single mode and boost the power on the laser. Typically, an extender will provide a single-mode cable distance of 30 km (18 miles).

Fibre Channel Disks

Fibre Channel disks have the highest capacity and transfer capability available. Typically, these disks have a capacity of 9GB and support redundant Fibre Channel loop interfaces.

Fibre-Optic Cable Connector

The SC connector is the standard connector for Fibre Channel fiber-optic cables. It is a push-pull connector and is favored over the ST connector. If the cable is pulled, the tip of the cable in the connector does not move out, resulting in loss of signal quality.

Gigabit Interface Converters

Distances in a data center are supported with twin-axial copper circuits, and therefore, hubs, disks, and many HBAs come standard with a copper interface. *Gigabit interface converters* (GBICs) and media interface converters plug into the copper interface and convert it to an optical interface. GBICs use an HSSD connector for the copper interface, and media interface converters use the DB-9 copper interface. The benefit is a low-cost copper link and optics for longer distance when required.

Gigabit Link Modules

Gigabit link modules (GLMs) are pluggable modules that provide either a copper or fiber-optic interface. GLMs include the *serializer-deserializer* (SERDES) and have a media-independent parallel interface to the HBA. Users can easily change the media interface from copper to fiber optics.

Host Bus Adapter (HBA)

An HBA is a card that fits into a computer, server, or mainframe that links it via Fibre Channel or SCSI to a storage device or storage network. HBAs are similar to SCSI HBAs and *network interface cards* (NICs). Fibre Channel HBAs are available for copper and optical media. A typical Fibre Channel *Peripheral Component Interconnect* (PCI) HBA is half-length and uses a highly integrated Fibre Channel *application-specific integrated circuit* (ASIC) for processing the Fibre Channel protocol and managing the *input/output* (I/O) with the host.

Hubs

Fibre Channel hubs are used to connect nodes in a loop. Logically, the hub is similar to a Token Ring hub with "ring in" and "ring out." Each port on a hub contains a *port bypass circuit* (PBC) to automatically open and close the loop. Hubs support hot insertion and removal from the loop. If an attached node is not operational, a hub will detect this and bypass the node. Typically, a hub has 7 to 10 ports and can be stacked to the maximum loop size of 127 ports.

Link Analyzer

Fibre Channel link analyzers capture the causes and effects of data errors. Specific frame headers can be monitored and captured for analysis.

Multimode Cable

Multimode cable is dominant for short distances of 2 km or less. Multimode has an inner diameter of 62.5 or 50 microns, enabling light to enter the cable in multiple modes, including straight and different angles. The many light beams tend to lose shape as they move down the cable. This loss of shape is called *dispersion* and limits the distance for multimode cable. The product of bandwidth and distance measures cable quality. Existing 62.5-micron *fiber distributed data interface* (FDDI) cable is usually rated at 100 or 200 MHz/km, providing gigabit communications for up to 100 or 200 m.

Routers: LAN Switch

Routers (LAN switches) interface Fibre Channel with legacy LANs. These are layer 2 and/or layer 3 devices that use Fibre Channel for a reliable gigabit backbone.

SCSI Bridge

Fibre Channel provides the ability to link existing SCSI-based storage and peripherals using a SCSI bridge. SCSI-based peripherals appear to the server or workstation as if they were connected directly on Fibre Channel.

Static Switches

Static switches, also called *link switches*, provide point-to-point connections and are controlled externally. They offer a low-cost option for applications not requiring the fast, dynamic switching capability inherent in the Fibre Channel protocol.

Switch WAN Extender

Fibre Channel switches can be connected over *wide area networks* (WANs) using an *interworking unit* (IWU). Expansion ports on switches are linked using either ATM or *Synchronous Transfer Mode* (STM) services. Since Fibre Channel may be faster than a single ATM or STM interface, multiple WAN channels can be used for full Fibre Channel bandwidths.

Addressing

Unlike many LAN technologies that use a fixed 6-byte MAC address, Fibre Channel uses a 3-byte address identifier that is dynamically assigned during login. N_Ports transmit frames from their own S_ID to a D_ID. Addresses in the range of hex'FFFFF0' to hex'FFFFFE' are special. Well-known addresses are used for such things as the fabric, alias server, and multicast server. Before fabric login, the N_Port's S_ID is undefined: hex'000000'. Hex'FFFFFF' is reserved for broadcast. In a point-to-point topology, fabric login will fail, of course, and the two ports simply will chose two unique addresses.

Arbitrated-loop devices still use the 3-byte address identifiers but also use an AL_PA. AL_PAs are 1-byte values dynamically assigned each time the loop is initialized. Once the loop is initialized and each L_Port has selected an AL_PA, public NL_Ports will attempt fabric login. If there is an FL_Port, the fabric will assign the upper 2 bytes of the NL_Port's address identifier and usually allow the low byte to be the NL_Port's AL_PA. (If not, the loop will need to be reinitialized so that the NL_Port can select the fabric-assigned AL_PA.) If no fabric exists, or if an NL_Port is a private NL_Port (does not log in with the fabric), the upper 2 bytes of the address identifier will remain '0000', and the lower byte simply will be the NL_Port's AL_PA.

However, there still needs to be a way to uniquely identify a port —even for much of the preceding initialization to take place. This is accomplished using name identifiers, a fixed 64-bit value. Name identifiers are used to uniquely identify nodes (Node_Name), a port (Port_Name), and a fabric (Fabric_Name). Name identifiers are not used to route frames but are used in mapping to ULPs.

Login

Fibre Channel defines two types of login procedures: fabric and N_Port. With the exception of private NL_Ports, all other node ports must attempt to log in with the fabric. This is typically done right after the link or the loop has been initialized. It consists of the node port transmitting a *fabric login* (FLOGI) frame to the well-known fabric address hex'FFFFFE'. The normal response is an *accept* (ACC) frame from the fabric back to the node port. Fabric login accomplishes the following things:

- It determines the presence or absence of a fabric.
- If a fabric is present, it provides a specific set of operating characteristics associated with the entire fabric, including which classes of service are supported.
- If a fabric is present, it optionally assigns or confirms the native N_Port identifier of the N_Port that initiated the login.
- If a fabric is not present, an ACC from an N_Port indicates that the requesting N_Port is attached in a point-to-point topology.
- If a fabric is present, it initializes the buffer-to-buffer credit.

Before an N_Port communicates with another N_Port, it must first perform N_Port login with that N_Port. Similar to fabric login, the process entails transmitting a PLOGI frame to the destination N_Port. Again, the normal response is an ACC frame. N_Port login accomplishes the following things:

- It provides a specific set of operating characteristics associated with the destination N_Port, including which classes of service are supported.
- It initializes the destination end-to-end credit.

- In point-to-point topology, it initializes the buffer-to-buffer credit.

Both fabric login and N_Port login are intended to be long-lived. Once logged in, a device can stay logged in indefinitely, even if it has no further data to transmit at that time.

Transmission Hierarchy

The easiest way to understand the methods by which information is transmitted over Fibre Channel is by looking at the form of a hierarchy.

8B/10B Transmission Character

At the lowest level, Fibre Channel uses the IBM 8B/10B encoding scheme. Basically, every byte of data that is to be transmitted is first converted into a 10-bit value called a *transmission character*. Using this encoding scheme has the following benefits:

- It improves the transmission characteristics of information to be transferred.
- It provides enough transitions to make clock recovery possible at the receiver.
- It improves the detection of single and multiple bit errors.
- Some transmission characters contain a unique bit pattern (comma) to aid in achieving word alignment.

Fibre Channel defines a 1 simply to be the state with more optical power (for optical links) or the state where the positive pin is more positive than the negative pin (in the case of copper). IBM 8B/10B encoding uses the idea of running disparity, which is concerned with the number of 1s and 0s in each transmission character. Running disparity is evaluated after the first 6 bits of each transmission character and after the last 4 bits. It can be positive (more 1s than 0s) or negative (more 0s than 1s). It is desirable to try to equalize the number of 1s and 0s over time. Every byte to be transmitted is encoded into one of two 10-bit representations depending on current running disparities.

Every byte to be transmitted is first converted into a 10-bit transmission character. There are many more possible 10-bit transmission characters than are needed to map to particular bytes. Most remaining 10-bit encodings are not defined, and only one is used. This is the special K28.5 transmission character that contains the comma—a 7-bit string that cannot occur in any data transmission character. The K28.5 transmission character is used as a special control character.

Transmission Word

All information in Fibre Channel is transmitted in groups of four transmission characters called *transmission words*. Some transmission words have K28.5 as the first transmission character. These are special transmission characters called *ordered sets*. Some ordered sets mark the beginnings and ends of frames (frame delimiters). Others convey information in between frames in the form of primitive signals (a single ordered set) and primitive sequences (a stream of the same ordered set). Examples of ordered sets are SOF, EOF, idle, receiver ready (R_RDY), LIP, ARB, OPN, CLS, and several others.

Frame

Fibre Channel defines a variable-length frame consisting of 36 bytes of overhead and up to 2,112 bytes of payload for a total maximum size of 2,148 bytes. The total size of the frame must be an even multiple of 4 bytes so that partial transmission words are not sent. Between 0 and 3 pad bytes are appended to the end of the payload to satisfy this rule. A SOF delimiter and EOF delimiter mark the beginning and end of each Fibre Channel frame. The CRC is the same 32-bit CRC used in FDDI.

Sequence

A Fibre Channel sequence is a series of one or more related frames transmitted unidirectionally from one port to another. All frames must be part of a sequence. Frames within the same sequence have the same SEQ_ID field in the header. The SEQ_CNT field identifies individual frames within a sequence. For each frame transmitted in a sequence, SEQ_CNT is incremented by 1. This provides a means for the recipient to arrange the frames in the order in which they were transmitted and to verify that all expected frames have been received. Multiple sequences to multiple ports may be active at a time.

Exchange

A Fibre Channel exchange is a series of one or more nonconcurrent sequences between two ports. The sequences may be in either direction. All sequences (and therefore all frames) must be part of an exchange. The originator of the exchange assigns the OX_ID field. The responder assigns the RX_ID field. As another perspective, one can use the following analogy:

frame = word

sequence = sentence

exchange = conversation

Of course, one main difference is that a Fibre Channel device can speak more than one sentence and hold more than one conversation at a time. *Information technology* (IT) managers, system integrators, and Value Added Resellers (VARs) quickly discover that Fibre Channel is built on the concepts and protocols they know well. Fibre Channel delivers the same types of functions, only faster, easier, more scalable, and much more reliable than SCSI and legacy networks.

Fibre Channel systems expand the flexibility of IT organizations with their inherent ability to run SCSI and IP protocols on the same network. These networks bring new levels of capability and perfor-

Fibre Channel

mance. Fibre Channel systems are built without restrictions. Virtually any topology that an IT organization requires is possible. The basic building blocks are point-to-point dedicated bandwidth, loop-shared bandwidth, and switched-scaled bandwidth. Switches and hubs are stackable. Fibre Channel networks and storage are built from products that are very familiar to IT professionals. Fibre Channel is what has made SANs a reality, and future developments on the interface likely will bring more features and faster speeds.

Summary

Fibre Channel attempts to combine the benefits of channel and network technologies. A channel is a closed, direct, structured, and predictable mechanism for transmitting data between relatively few entities. Channels are used commonly to connect peripheral devices such as disk drives, printers, tape drives, and so on to workstations. Common channel protocols are SCSIs and HIPPIs.

Fibre Channel is the preferred technology for transmitting data between computer devices at a rate of up to 1 Gbps (1 billion bits per second). It is especially suited for connecting computer servers to shared storage devices and for interconnecting storage controllers and drives.

Fibre Channel hubs are used to connect nodes in a loop. Logically, a hub is similar to a Token Ring hub with "ring in" and "ring out." Each port on a hub contains a PBC to automatically open and close the loop. Hubs support hot insertion and removal from the loop.

Fibre Channel defines three topologies: point to point, arbitrated loop, and fabric. Before a loop is usable, it must be initialized so that each port obtains an AL_PA, a dynamically assigned value by which the ports communicate. It maps to the lowest byte of the native address identifiers (_ID and SI_ID). Although the AL_PA is 1 byte long, only 127 values are valid (neutral running disparity). If more than 127 devices are present on the loop, some will not be able to select an AL_PA and will be forced into nonparticipating mode.

The concept of flow control deals with problems where devices receive frames faster than they can process them. When this happens, the result is that the device is forced to drop some of the frames. Fibre Channel has a built-in flow-control solution for this problem. A device can transmit frames to another device only when the other device is ready to accept them. Before the devices send data to one another, they must log in to one another.

Fibre Channel defines several communication strategies called *classes of service*. The class used depends on the type of data to be transmitted. The major difference among the classes is the types of flow control used.

Fibre Channel arbitrated-loop configurations consist of several components, including servers, storage devices, and a Fibre Channel switch or hub. Another component that might be found in an arbitrated loop is a Fibre Channel-to-SCSI bridge, which allows SCSI-based devices to connect into the Fibre Channel-based storage network. This not only preserves the usefulness of SCSI devices but also does it in such a way that several SCSI devices can connect to a server through a single I/O port on the server. This is accomplished through the use of a Fibre Channel HBA. The HBA is actually a Fibre Channel port. The Fibre Channel-to-SCSI Bridge multiplexes several SCSI devices through one HBA.

Chapter 6 presents case studies of SANs implementations.

CHAPTER 6

Case Studies

This chapter is a compilation of case studies intended to provide you with insight and information about the experiences of other companies that have implemented the *storage area network* (SAN) solution. This is not presented as an endorsement of any particular storage company. Grateful acknowledgment is extended to *Storage Networking World On-Line* (SNW) for allowing us to reprint from their archives. SNW is a publication that is dedicated to "providing timely, useful information for planning and implementing high-end storage solutions and services."

SAN Implementation Saves Money and Boosts Network Performance at New York Life

A bottleneck problem and towering storage costs led to a happy, although complex, solution at New York Life, a Manhattan-based mutual insurance company with $84 billion in assets.

In fact, since moving from direct-attached storage technology to a SAN setup from EMC in Hopkinton, MA, New York Life has saved "a couple of million dollars" on storage equipment costs while also boosting network performance and availability, according to Michael Polito, a corporate vice president responsible for enterprise storage and business resumption services at the large insurer. In addition, the company has gained the ability to better monitor its network, which enables it to plan future server deployments more accurately.

However, although the move from direct-attached storage to a mesh network of 90 servers and 12 switches has spurred network performance, it also has increased *information technology* (IT) complexity, a situation that New York Life is moving rapidly to resolve.

In the Beginning . . .

It was the end of 1999 when New York Life concluded it had to reduce storage costs by implementing a SAN. Its direct-connect stor-

Case Studies

age infrastructure had become uncontrollably expensive, and the following spring, a storage assessment substantiated this conclusion. "We felt that going with SAN technology was a better way to limit costs," Polito explains.

The company's previous infrastructure consisted of EMC Symmetrix devices attached directly to a mixed bag of 90 servers. The server platforms ranged from Compaq boxes running Microsoft Windows NT, to Sun Microsystems servers running Solaris, to IBM RISC devices running AIX. Most servers were dual-attached to a Symmetrix box for reliability and redundancy. And since each Symmetrix device had 32 connectivity ports, each could support a maximum of 16 servers.

However, the 32-port limitation was a bottleneck. Even though the Symmetrix devices could have supported a lot more disk capacity than 16 servers typically require, there was no way to add a seventeenth server to make use of that capacity.

"So we wound up paying for a new Symmetrix device to support additional servers and were unable to leverage unused DASD capacity on the other device by linking it to additional servers," Polito says. In addition, IT mirrors corporate data in order to make them as highly available as possible. This replication, although valuable, greatly increased storage administrators' data maintenance burdens, and it used up more capacity as well.

The Solution

Last November, New York Life tackled the problem by implementing eight EMC Connectrix Fibre Channel switches, which provide any-to-any connectivity between the servers and Symmetrix devices. This has freed up captive storage capacity to feed the growing data needs of new servers and applications.

The initial SAN installation consisted of two fabrics of four switches each. Earlier this year, in order to accommodate escalating traffic switching needs, the insurer added two switches to each fabric, for a total of 12. Servers are connected to both fabrics to ensure continuing availability if one fabric goes down.

Saving Big Bucks

The effect of the SAN on New York Life's storage costs has been significant and immediate. "We haven't had to add a new Symmetrix box on our floor for a year," Polito declares. "Before the SAN initiative, we faced a need to bring in three or four new Symmetrix boxes. Not having to buy those devices has saved us a couple of million dollars."

In addition, it has become more manageable and less disruptive to attach new servers onto the SAN or put new DASD into existing Symmetrix frames.

The SAN implementation has done more than save money on storage equipment costs, however. The Connectrix fabric also has become the foundation for several strategic initiatives aimed at reducing storage costs and ensuring reliability and performance levels on a proactive, ongoing basis.

A By-Product: Complexity

Getting a handle on storage use and traffic patterns is a prerequisite to effectively maintaining the SAN infrastructure. However, this is no easy task in New York Life's complex SAN environment. For instance, while SANs make it possible to ensure better performance and fault tolerance for applications by distributing shared databases across multiple volumes, those capabilities also make it a lot more difficult to design applications.

"It isn't a discrete process. It's an ongoing evaluation process," Polito notes.

The company's IT department wants to be able to comprehensively model storage demands and traffic patterns—not only for individual applications and servers but also for the aggregate activity of all those servers going through all those switches. This will help storage managers make decisions such as which Connectrix ports to plug servers into, where to put additional servers, and how to optimize performance by understanding zone requirements.

The goal is to identify and address bottlenecks—such as those created when too many servers are funneled through one Connectrix switch—before performance degrades to the point where users start complaining.

New York Life is still looking to avoid performance problems by finding modeling tools that can "crunch" data and determine the correct zoning for applications and servers. "We're starting to see products from companies like Veritas and EMC," Polito comments. "With SANs only a few years old, it's not surprising that the SAN management market has some growing up to do."

Analyzing Storage Needs

Polito's group also has been implementing a formal methodology for analyzing the current and projected storage needs of end users and in-house developers. "We're actively working on a storage demand project, which will put an infrastructure in place to capture utilization and report back to business units on what they're using and the costs involved," he says.

One goal of the project is to accurately tie the deployment of storage capacity more closely to real needs. "When a user asks for 100 GB of storage, we'll ask, 'Do you need it right away, or is this a projected need over the next six months?' If an application's storage demands had been projected to grow 80 percent over this time period, and they only grew 3 percent, we're going to ask why," Polito states. Toward this end, IT hopes to institute a charge-back program for storage in the near future.

The Value of an Independent Consultant

In designing these initiatives, Polito and his group turned to third parties for help. It hired consultancy ITIS Services in Norwalk, CT, as well as EMC, to assess, justify, and propose a design for the SAN installation. "ITIS Services brought a broad view of SAN technology to New York Life," Polito says. "Whether testing products in their own laboratory or helping companies design and implement storage network infrastructures, they talk from experience."

ITIS technicians spent 10 days locating unused captive storage on various Symmetrix devices. According to Brendan Reilly, CTO at ITIS, that storage could be allocated via the SAN fabric to whatever server or application needed it. As a result of ITIS's efforts, New York Life was able to purchase 4 TB less of disk space than EMC originally recommended.

Reilly says that ITIS encourages clients to take full advantage of scalable SAN architectures and their hot-swapping capabilities to add storage capacity on an as-needed basis. "Gartner says storage costs are coming down 44 percent annually, so if you defer purchase of storage by 6 months, you save 22 percent of the cost of those next terabytes," he explains.

Polito hopes that a more accurate and comprehensive view of storage use and traffic patterns will help educate both end users and business managers. "If people understand how applications are designed, managed, and used, they will be more appreciative of the cost of the resources they require," he says. This in turn will make it easier for IT to justify purchases of storage, SANs, and related leading-edge products in the future.

Upper management already has enough data to ensure its support of these cost-containment initiatives. "When we shot from 4 to 19 TB in a year," Polito recalls, "that was a real eye-opener."

R&B Group Adds Redundancy and Speed with Fibre Channel Adapter Cards

In the winter of 2000, R&B Group, a graphic image and photography company that offers state-of-the-art imaging and printing services, undertook the project of converting its cumbersome data storage system into a heterogeneous SAN.

The SAN, which provides a high-speed data storage network for the entire organization, was created to enhance R&B Group's connectivity, provide dual multipathing redundancy, and, most important, deliver critically needed additional bandwidth for downloading large files.

Overcoming Major Drawbacks

"Our old system had two major drawbacks," says Dan Tesch, director of technology at R&B Group. "We had no redundancy for our data because it was not RAID and it was extremely slow when it came to downloading large files."

R&B Group's SAN is located at their production facility in Chicago. It contains 1 TB of storage and houses all the company's production data, which consist of very large Adobe Photoshop, Quark Express, and EPS files. The SAN deals strictly with the company's stored and ongoing graphics projects.

With the assistance of Wenzel Data and Bell Micro, the R&B team carefully identified and evaluated best-of-breed Fibre Channel technologies they would use throughout the system. On recommendation, they decided to go with JNI Corporation's FCE 3210-C Fibre Channel adapter cards.

"There really wasn't much to think about when it came to the decision about JNI's Fibre Channel adapter cards," Tesch comments. "We required heterogeneous support, and JNI seemed to be the best game in town."

The JNI cards joined a technology lineup that included Hewlett-Packard servers, Windows NT, Hitachi DK32CJ-36FC drives, and a Mylex RAID controller.

"Getting the JNI cards up and running was a plug-and-play scenario," declares Tesch. "There were no configuration problems, our defaults were instantly accepted, and the system has been working fine ever since."

Improved System Access

Since its implementation, the SAN has been a huge success. R&B employees can access data quickly from anywhere in the company, and the storage system itself, due to the dual multipathing and redundancy, is more stable.

"The biggest and most important asset of this new system is the speed," says Tesch. "We deal with huge graphics files—some up to 1.2 GB—and this new system is three to five times faster than our previous system."

Files that used to take half an hour to download now take only about 10 to 12 minutes, says Tesch, adding, "With this new system, we are no longer slaves to long transfer times."

Because all the storage is shared in a SAN, R&B Group no longer has to purchase individual storage units for each server. Now it just has to buy enough storage to support its capacity needs for the entire server farm.

"We can also rest assured that our data system is more stable," says Tesch. "The dual multipathing data are saved on two separate disks that are connected to separate channels. So when one goes down, retrieval from another is not a problem."

Forest Products Manufacturer Takes Control of Its Storage

Boise Cascade Corp., based in Boise, ID, ranks as the country's largest integrated manufacturer and distributor of paper and wood products. With $8 billion in 2000 revenues, it is also a major U.S. distributor of office products and building materials. Boise Cascade owns and manages more than 2 million acres of timberland in the United States.

In order to gain more control over its storage resources, the company recently purchased StorageCentral SRM from WQuinn, Inc., which was acquired recently by Precise Software Solutions in Westwood, MA. StorageCentral SRM is a policy-based storage resource and performance management software package.

The software paid for itself right out of the box. It helped reduce the company's primary file server's full backup window by 30 percent and helped the company reclaim about 25 percent of the wasted space on that server. These two events helped the company to better prepare for moving its files to the new Magnitude SAN system from XIOtech Corporation in Eden Prairie, MN. Going forward, StorageCentral SRM is going to help the company keep the cost of this very active file server under control as it pays for only the SAN space the company needs.

Until recently, about 1,200 employees in the company's southern Minnesota plant stored their home directories on a Gateway file server running Windows NT. The server—which maintained 70 GB of data—also houses some shared departmental directories and one directory used by the entire plant. The problem was that the window for a full server backup had grown to about 24 hours.

Before the company could even think of moving any files to the SAN, management felt that it was necessary to audit the contents of this server. The company did this by running StorageCentral SRM reports that describe the characteristics of files. These characteristics include duplicate reports, files more than 1 year old, files by media type, and usage by share. This information enabled the company to pinpoint pockets of wasted space. It was discovered that most of the employees had been dumping every file they had to their home directories for possible future use.

Cleanup Campaign

The company immediately launched a cleanup campaign, telling everyone that the company would be allocating 150 MB for home directory space and holding everyone to his or her space allocation. The company added that employees would have a month to go through their files and get rid of what they did not need or have extra files burned to CDs.

If a home directory exceeded the 150 MB amount, the StorageCentral SRM was used to send the employee a report via an Excel spreadsheet listing his or her files.

Making sure employees didn't move their files into the departmental shared folders became a key concern to the company. Once a week the company ran StorageCentral SRM reports to see if the size of the home directories decreased or the size of the shared folders increased. Everyone cooperated. As a result, the help desk phone didn't get many calls, and employees who truly needed more space got it.

Within a month, this freed up about 20 GB of wasted space. Overall, for every employee whose files was archived, there were six employees who found that they did not need to keep extra files.

Backup Time Savings

Reducing this amount of space enabled the company to chop 8 hours off its backup window. As a result, the company can now do a full backup on Friday night rather than waiting for the weekend. During the week, the company backs up only the changes since the most recent backup.

Next, the company used StorageCentral SRM's ability to monitor the amount of space everyone used. Now, as employees reach 90 percent of their space allocation, StorageCentral SRM sends them a message saying how much space they have left and how they can free up space. If employees hit their space limit, they are allowed to save what they are working on, but they must either free up space or call the help desk. Employees understand the reason they need to stay within their limit and take the appropriate measures to do so.

In an effort to maintain good file management practices and shrink the company's backup window even further, the company will continue to use StorageCentral SRM on the SAN. In fact, since the tape library will be attached directly to the SAN, the company can back up files faster than it could if the files were backed up across the network.

The company also plans to use the same type of home directory space allocation and monitoring procedures on the SAN. In so doing, it plans to install similar procedures for shared departmental folders. By running StorageCentral SRM reports across each shared departmental folder, the company can easily determine how much space it should allocate to each folder. The company also plans to have the alerts go to the employees who are responsible for each folder.

On another front, StorageCentral SRM's file blocking feature will help the company automatically enforce its policies relating to where employees can store their files. For example, employees are not supposed to store digital images in their home directories but rather in shared departmental folders. With StorageCentral SRM, the company can set filters on certain media file types and designate the areas that are to be blocked from storing these file types.

Alternatively, the company can designate areas that can accept these file types. If employees try to save a blocked file type in their home directory, they will get a warning message describing why they cannot store that file.

StorageCentral SRM has been very successful at Boise Cascade so far, and the company plans to gain more and more benefits from it in the future.

Two Risk-Averse Users Tread Cautiously During SAN Development

Do you think SANs are really ready for prime time? Do you believe that they are universally interoperable? Is it your opinion that there are plenty of multivendor mature management tools out there? After listening to two prominent users speak at Storage Networking World last week, the only plausible answer to these questions is no.

On the other hand, are some SANs already in place and rapidly reducing their total cost of ownership? Can savvy users at least take the first steps toward integrated storage networking environments by forcing their vendors to sit down together and map out integration strategies? Is the future bright with potential SAN benefits? This time, the two speakers would say, "You bet."

The two users in question are Gary Fox, senior vice president and director of Enterprise Data Storage at First Union Bank, and Michael Butler, vice president of Morgan Stanley Dean Witter. Both have been in the SAN trenches, both have felt the pain of implementation, and now both are increasingly enjoying the fruits of their labors.

Gary Fox: Developing a Nationwide SAN

Fox is developing a nationwide SAN while riding the rails of 24/7 uptime requirements. The impetus for this SAN was First Union

Bank's *Enterprise Check Image Archive* (ECIA), which was created to replace the bank's outdated microfilm data access system. Each day, the First Union SAN has to handle some 600 GB of incoming data consisting of 17 million images that average 35 KB in volume. It also has to deliver data to desktops in 5 seconds from disk and 80 seconds from tape. On top of this, the bank is required by the *Federal Deposit Insurance Corporation* (FDIC) to back up each of its 5,000 servers.

Building the infrastructure required to meet all these specs is no easy job. As Fox puts it, "Network architecture is a hard thing to come by."

Fox hears the vendor rhetoric about SAN compatibility, but the harsh realities of his job have driven him to keep his SAN environment as simple as possible. In his opinion, a homogeneous switch infrastructure is the safest way to go. He also says that the most notable weak points he has to deal with are *host bus adapters* (HBAs) and disk and tape drivers, saying that most of his SAN problems to date can be traced back to these areas.

The further Fox got into SAN development, the more he realized that he would have to rely on internal resources and a focus on heterogeneous technology. Even though he considered using a storage service provider, he did not go that route because his SAN environment was too highly customized.

"My advice is to standardize, particularly on HBAs," he declares. "And be prepared to spend a lot of money on the fiber architecture."

Michael Butler: Seeking Business Continuity

With 250 TB of institutional storage and 4000 Microsoft Windows NT and UNIX servers accessing that data in a 24/7 environment, Morgan Stanley Dean Witter is focused primarily on risk aversion and business continuity. According to Butler, these are the two primary reasons why the financial institution built five SANs in the metropolitan New York area. Other reasons include the need for reduced time to market and the desire to take advantage of best-of-breed technologies.

In order to satisfy his company's SAN requirements, Butler and his team came up with the concept of "campus computing SANs" based on host clusters, network clusters, and storage clusters. In the course of developing and deploying this SAN strategy over a year ago, he faced several challenges, including a lack of management tools, a dearth of vendor support, and a nearly complete absence of interoperability.

Despite this gloomy technological landscape, the highly risk-averse Butler had to develop and integrate a 24/7 system that would handle billions of dollars each day. And he had to do this while dealing with a mish-mash of vendor-specific HBA switches and host-specific fail-over software. Not a pretty picture.

So how did he do it? "We became our own integrator," Butler says. "It wasn't the business we wanted to be in, but it was the best tactical choice." And how did he deal with the dearth of management tools? Same strategy. "We developed the tools in house, and we told our vendors, 'You adhere to our guidelines or you are out of our SAN.'"

It has been a battle, but Morgan Stanley Dean Witter is slowly achieving the SAN homogeneity that is increasingly allowing Butler to view storage as a commodity. As for time to market, the financial services institution can now add new servers "instantly," although the risk-averse Butler still prefers to do so overnight. And after previously restricting his servers to 50 percent capacity tops, he has relaxed to the point where he now lets them run at 70 percent.

Speaking of his SANs, he declares, "They've been rock solid. They really do work as advertised. They are solving real problems for us."

Law Firm Backs Up Data to "EVaults"

We've all heard it before—lawyers live and die by the clock. Working long hours to meet crucial deadlines, there is little time for mistakes, especially those involving the accidental deletion of crucial documents. Unfortunately, these mistakes do happen, especially in law firms, where so many documents are created.

Resorting to tape backup helps retrieve lost data, but it can be a time-consuming and expensive process.

One law firm, *Fraser Millner Casgrain* (FMC), has found a way around this monumental time and resource drain. FMC is one of the leading Canadian law practices, with over 1,500 employees—550 are lawyers—spread out across locations in Toronto, Montreal, Ottawa, Calgary, Edmonton, and Vancouver.

Each of the lawyers in these locations shares important information and files, which can lead to problems. According to Lewis Robbins, chief information officer of FMC, "We have to back up on a daily basis. Someone will either accidentally delete or overwrite a document, which means that we need to go back and attempt to retrieve the data in a timely manner. This happens about six times a week."

The key word in that sentence, according to Robbins, is *timely*. Because lawyers work with critical, sensitive information that can affect the outcomes of major cases or lawsuits—and are generally strapped for time—documents must be retrieved as soon as possible. This was a difficult task with FMC's original tape backup system.

"There were two major problems that we were running into," says Robbins. "First, we were spending a lot of money on backup tapes. With the amount of use we were getting out of them, we were faced with the continual purchase of new tapes, as well as making a significant investment in multiple tape loaders. This was getting to be very, very expensive."

Second, Robbins continues, the tape-based system proved to be very time-consuming. "Whenever we wanted to retrieve a document that had been accidentally lost, it took, at times, up to a whole day," he says. This is so because someone had to physically retrieve the tape, bring it back to the office, and then search for the needed information. "That takes a lot of effort and man-hours, and waiting for the necessary information to be retrieved is a very painful process for lawyers," he says.

The Solution: EVault's Online Backup

When tape backup became too costly and time-consuming, Robbins turned to a company that was providing a useful alternative to tape-based systems. EVault, a provider of online data backup and disaster-recovery solutions, presented Robbins with a plan to transfer all of FMC's data to secure off-site vaults. This process—known as "EVaulting"—was designed specifically to easily back up data at any preset time of the day and automatically and safely transfer them using the Internet. Once the data are secured, companies can easily and quickly access the data they need, performing a full restore in a matter of minutes rather than days.

"With EVault, we're able to back up data online every night, after the work day is over and no one else is around," says Robbins. "Then, once we do need to retrieve information, we can access it online and get it back and fully restored in about a half-hour—a far cry from the days when it would take 24 hours to get everything back in order."

Robbins says that he is using EVault primarily to manage backups across a mixed Microsoft and Novell network in two locations—Toronto and Ottawa. "There are about 600 employees spread between the two of them," he says, "which means a lot of data to manage."

So far, so good. Robbins says he is happy with the time EVault has saved him and his colleagues. "If you figure that FMC has a couple of million documents online and over 200 GB of data, you have to figure that some U.S. law firms hold substantially more," he declares. "That's what makes EVault's online data backup so important for law firms. The true *return on investment* (ROI) here is the speed with which you can use EVault to recover data, regardless of how much information you are backing up. It's much, much faster and more reliable than tape."

Global Provider Relies On Flexible Storage

As an *application service provider* (ASP) and an IT consultancy, TDS Informationstechnologie AG faces all the challenges that modern e-business can present. Headquartered in Neckarsulm, near Heilbronn, Germany, and with revenue of 160 million euros last year ($138.8 million), the company's consulting services encompass all tasks involved in the setup and operation of company-wide IT solutions.

In the ASP side of the business, clients can connect to TDS's network to access their data and applications, which are stored at the company's computer center—one of the largest, most modern computer centers in Germany, with 530 servers. There are several powerful benefits: IT costs are transparent; initial investments in software licenses, hardware, and know-how become obsolete; and TDS continues to be the owner and license holder of the software used. Moreover, applications are managed efficiently from a central location, and users receive support that is tailored to their environment. The bottom line for clients: the opportunity to focus fully on their core business.

However, with its ever-more-complex requirements and increased information flow—not to mention the growth of its successful hosting business—TDS realized last year that it needed a new solution for the use of data storage.

The Challenge: High-Availability Storage

"It's our goal to utilize and pay for storage capacity in tune with our customers' needs," says Thomas Gebhardt, TDS division manager for application hosting and a member of the top management team. In addition to ensuring high-availability storage capacity on demand, TDS insisted that any storage solution had to provide 7.2 TB of hard-disk capacity. This level of storage is required for optimal performance, due to the structure and size of the SAP databases with which the company frequently works.

After intensive market research, TDS opted for the concept of storage on demand. The conception, planning, design, and implementation of the storage system started in March 2001.

"The solutions offered are mostly very complex and require a high degree of investment, as well as expenditures for adaptation and integration," Gebhardt says. "We chose storage on demand because this solution combines an optimal cost-benefit ratio, with high-availability storage capacity, easy integration into existing business processes, and more efficient adaptation into future requirements."

The Solution: Storage on Demand

Storage on demand combines demand-oriented costs with efficient management of disk and tape storage. As with monthly electric bills, customers can receive their individual storage requirements from a "virtual socket." They pay for only what they actually use in storage volume. For TDS, this meant that for a flat monthly fee, a systems integrator would provide the IT service provider with on-site storage capacity in exactly the required quality and quantity.

In less than 6 weeks, the new TDS storage solution was complete. Operation started with an initial capacity of 500 GB, for an initial run of 36 months.

TDS's data backup is handled by Legato Systems' NetWorker. Every day, the software saves a data volume of 5.2 TB. NetWorker 6.0 offers a tailor-made solution not only for TDS but also for all companies that back up vital business data internally, as well as for service providers that offer data security services based on service-level agreements. The Legato solution assured TDS that it would be supplied with the latest technology at the current time and in the future as well.

Legato NetWorker modules supplement NetWorker 6.0 to provide high application availability and high productivity of the database administrator. The modules increase application availability by reducing and often totally eliminating the need to shut down databases or applications. Quick and powerful online backup-and-restore functionalities make important data available from databases and applications.

Important Criteria

The pivotal criteria for TDS's decision to use NetWorker 6.0 were its manageability, performance, and scalability—extremely important factors for TDS's numerous SAP databases. The NetWorker module for SAP R/3 under Oracle assumes the role of a scalable high-performance backup system, securing integrity, and high availability of TDS's SAP R/3 data in Oracle databases under both Windows NT and UNIX. The data are protected on both the database level and the transaction log level.

The hardware selected to support the storage-on-demand system environment was EMC's Symmetrix 3930. TDS's previous experience with similar technology had been positive in terms of functionality, performance, maintenance, administration, and configuration options. The server systems are connected using a switched fabric, which is an integral component of the solution. Certain parts of the storage system are physically mirrored with Fibre Channel and *Symmetrix Remote Data Facility* (SRDF).

TDS is now able to react flexibly to the respective market offerings of servers because EMC has certified leading suppliers' servers for connection to EMC storage systems. This takes into consideration all the demands that TDS end users can make in terms of possible server systems. Furthermore, servers can be selected and implemented flexibly, depending on technical and commercial criteria.

Within the framework of the storage-on-demand model, existing systems are integrated into the solutions via upgrades. EMC Symmetrix disk reallocation enables hard-disk capacities to be assigned flexibly. Server access to the assigned storage areas is controlled by Volume Logix, which protects access to hard-disk capacities and secure control. This capability is particularly applicable to a mixed connection of servers under UNIX, Windows NT, and Windows 2000.

By combining EMC's software solutions, TDS is able to authorize hard-disk capacities quickly and easily, as well as assign disk capacities to servers connected by Fibre Channel switches by Brocade. The project-related work for acquiring and implementing storage components is therefore reduced to a minimum.

The Benefits: Economy, Flexibility

The storage-on-demand technical facilities can be almost indefinitely expanded to include additional servers and storage systems, which can be connected using implemented ESN/SAN components, Fibre Channel connections, Fibre Channel switches, and software solutions. The solution thereby accounts for the growing demands placed on availability and security today and in the future.

Case Studies

The solution's success comes not only from powerful technology but also from commercial considerations. This is the only way that modern storage on demand will function.

According to Karl-Heinz Fetzer, TDS's director of hosting services, "We have decided on a solution with the economically most feasible components from the most flexible suppliers."

Data Restoration Package One-Ups Lovebug and Anna Kournikova

When the Lovebug computer virus struck genetics firm Orchid BioSciences in Princeton, NJ, early last year, gigabytes worth of e-mail information were corrupted, and the company had to scramble to keep its business running.

This was a disaster because Orchid's main source of income is performing DNA analyses, and many of the technical discussions between scientists were contained in the e-mail files that the virus corrupted. To make matters worse, the May 2000 virus attack came on the same day the firm was holding its initial public stock offering.

"It made for a real nightmare," says Phil Magee, Orchid's director of IT. "I had to bring in a lot of consultants. We pulled the plug on the Exchange server and restored most of the data from tape backups."

However, Magee learned his lesson. If the rapidly growing Orchid was to avoid another bout with computer viruses, it needed a better way to back up and restore e-mail that contained vital technical information.

Here's why: Orchid does genetic analysis work for pharmaceutical and agricultural companies and for university researchers by unraveling the double helix of DNA. It also operates medical laboratories that do disease diagnostic work for doctors, determine child paternity, and test for bone marrow compatibility for bone marrow transplant patients. To do this work, Orchid's scientists rely heavily on e-mail to exchange technical data, a habit that is not likely to change just because computer viruses pose a threat to e-mail. Because Orchid uses the high-level Triple *Data Encryption Standard* (DES) e-mail encryption, "Our people are not reluctant to put vital technical information in e-mail," Magee says.

To make matters more difficult, the virus risk goes beyond the loss of valuable technical information contained in e-mail. "We use Exchange not just as e-mail but as groupware," Magee says. "There are public folders, address books, and calendars. I can set up an appointment in Princeton, invite attendees from all the locations, and have it pop up on their calendars that they are supposed to be here. I also can reserve a conference room by 'inviting' it to the meeting."

In a company so reliant on the Exchange server, even a single corrupted piece of data could cause big trouble, Magee says. For example, business would be disrupted if a virus corrupted data and eliminated a reservation for a conference room.

As a result, Orchid went looking for a data-restoration product that had the most attractive means of handling Microsoft's Exchange e-mail server. After trying a total of six backup products, Magee chose the Galaxy product from CommVault Systems, Inc., of Oceanport, NJ.

One of the principal benefits was CommVault's ability to restore small sections of the Exchange database in a precise way, an operation that solved many data-corruption problems but which took less time than restoring the entire multigigabyte Exchange database, Magee says.

John Barry, CommVault's director of product management, says the software was designed to retrieve the smallest piece of information that a server handles. In the case of Exchange, this is an e-mail message.

Magee says CommVault's limited restoration of an Exchange database is particularly useful when Orchid is trying to restore its e-mail system after a virus attack. Because Orchid's IT department usually can determine the exact time the virus struck and the exact types of destructive files in which it arrived, it is possible to use the CommVault software to restore only the data that were changed after the virus struck and to avoid restoring files of the type that contained the virus.

"If you can restore just that particular day or week's worth of data, it makes for a speedy recovery," Magee says.

Case Studies

This ability to recover small chunks of data quickly rather than performing a lengthy database restoration is particularly important to Orchid, which at any given time has 8 to 12 GB of data on its Exchange mail server. A full restoration of this much data would take 10 to 12 hours, Magee says. However, restoring the e-mail box of a single user, including calendar and other nonmail items, takes only about 10 minutes with CommVault's Galaxy, he says. System administrators handle the recoveries because individual users are not authorized to do so.

The software also saved time in other ways. Unlike some backup and recovery products, Galaxy restored corrupted messages not only to the right e-mail box, but also to the right portion of the e-mail box, such as the inbox, outbox, drafts, or deleted items folders, Magee says. It also correctly restored items to the calendar and the contact list.

"Most of the other products I tried for restoration didn't work satisfactorily because they either put the data in the wrong location or recovered the data incompletely," Magee says.

In addition, the software is easy to use. The server-based package uses a web browser interface that makes it practical for IT people to recover data from any user's workstation. Recovery products from other firms did not have such a full-featured interface, he says.

At the moment, Orchid is still rolling out Galaxy software to several company locations and is able to use it to restore e-mail for about 350 of the firm's 500 employees. The firm's main locations are in Princeton, Dallas, Dayton, Sacramento, and the United Kingdom.

Dealing with the Anna Kournikova Virus

The firm's determination to extend Galaxy to all employees is based on its second experience with a computer virus. When the Anna Kournikova virus struck Orchid's Exchange servers earlier this year, just weeks after Galaxy was installed, it highlighted the ease with which the package handled data restoration, Magee says. This was in sharp contrast to the much greater difficulty Orchid had in restoring a second Exchange server that, while it contained a much smaller amount of data, was not using Galaxy.

The Anna Kournikova virus caused the Exchange databases to balloon in size by sending additional e-mail to people in the address book of infected computer workstations. The affected Exchange e-mail server that was using Galaxy had a total of 6 to 7 GB of data, and Magee says having to do a complete restore would have been an arduous process. In addition to deleting the new virus-bearing messages, it would be necessary to defrag the disks to get rid of the blank spaces where the deleted messages had been. What's more, Magee's IT staff would then have had to search for corrupted data.

Using traditional Microsoft techniques to accomplish these tasks would have required taking the Exchange server offline for 10 hours, Magee says. Instead, he used Galaxy to restore the server to the content it had contained a day earlier, before the virus struck. By restoring only items that had changed after the virus struck—and using file filters to avoid restoring virus-containing messages at all—the work was completed in 3 hours.

Meanwhile, the Orchid IT department spent 6 hours restoring a second Exchange server that was not using Galaxy, even though the server contained only 1 GB of data. In addition, more time had to be spent searching for corrupted files and defragging the disk.

"The pains we went through to restore that 1 GB database were the big eye-opener for us," Magee says.

Euroconex Turns to MTI Vivant for High-Availability Data Storage

Nova, the third largest credit card processor in the United States, has teamed up with the Bank of Ireland to create a new joint-venture company called Euroconex. The company aims at offering end-to-end credit card transaction processing to acquiring banks, institutions, and merchants in Ireland and the United Kingdom. Euroconex plans to expand throughout Europe, where outsourcing of the retail payment process has so far been limited.

Euroconex will operate from a new facility located about 40 miles south of Dublin, in Arklow, where back-office activities were due to begin this spring with 70 of the Bank of Ireland's staff joining the new operation. These customer service activities will rely on an entirely new IT system, custom designed with the help of Datalink, a U.S.-based network integration consultancy.

The system will rely primarily on a Microsoft Windows 2000 active directory infrastructure. However, Euroconex also intends to use Oracle and Sun Microsystems' Java technology for its merchant accounting database. Thus, it requires a data storage system capable of servicing multiple platforms.

The Solution: MTI Vivant

The company turned to MTI and its Vivant product to provide multiplatform high-availability data storage. "Nobody else did Windows 2000 with a SAN on the back end," says Matthew Yeager, senior systems architect at Datalink.

Nova had already chosen an MTI system over Hitachi Data Systems and EMC in the United States when its SAN in Atlanta was in need of more capacity. "We had already learned a lot from the system implementation in the States, and we wanted to take what we had learned from this project and bring it with us to the Irish development," says Gene Budd, director of technical support at Euroconex.

"When we sourced a storage vendor in the States, we looked at all the main players and asked them to test the products over a 2-month period. MTI not only did what it said it would, but it also provided us with the most cost-effective solution. EMC and MTI were very close speedwise, but we found MTI's management interface a lot simpler. It was both a management interface issue and a cost issue."

Euroconex needed a solution that would be scalable and easy to manage. It decided to base the systems infrastructure around MTI's Vivant 20, which provides a storage capacity of 1.5 TB, upgradeable to 4.5 TB with the adjunction of more disks.

The system is fault-tolerant, thanks to two Microsoft Exchange 2000 high-availability nodes implemented in an active-active clustered configuration (meaning that they are both alive, as opposed to having one node waiting for the other one to fail). Each of these nodes is capable of handling its counterpart's burden should there be a failure. They are connected to the Vivant 20 through HBA Fibre

Channel cards. This configuration also enables data to be fully accessible during maintenance operations through the use of rolling-upgrade techniques.

Disaster Recovery

Disaster recovery and backup also are provided by MTI, using the new version of Legato backup software on ATL tapes for off-site storage. "We did some testing of the disaster-recovery element of the system on a temporary system recently, and we can be up and running again, with full Windows content, within 20 minutes," Budd explains.

Euroconex wanted its users to be able to access their data no matter which workstation they were using on any day. This has been achieved by including user directory folders in the user profiles stored in the active directory. These data are then stored directly on the Vivant 20. Moreover, all Microsoft Office documents are saved automatically to the user directory located on the server.

Achieving High Availability

In addition, Exchange 2000 is able to implement separate databases for individual user groups and therefore achieve higher availability —as opposed to having a single contiguous database for all users. Nine separate logical unit numbers thus have been created on the Vivant 20, where all data resources will be centralized.

The infrastructure will be used for all customer support activities and will enable storage of call monitoring recordings and imaging documents such as contracts or screen grabs.

Implementation of the system began in November 2000 in a test center located in Sandyford, Dublin, 7 months after design analysis work first started. According to Budd, the Euroconex experience will now benefit Nova's U.S. operation and enable the company to collapse one of its data centers there. "The clustering technology built into Exchange 2000 will allow one clustered server to serve multiple sites without the worry of having one system down bringing the whole e-mail system to a halt," Budd says.

The joint venture will employ approximately 800 people, mostly in new positions, within the next 4 years. The project was scheduled to go live on June 1, when Euroconex expected to start processing merchants on its own system.

With $100 Trillion in the Balance, DTCC Plays It Close to the Vest

If there is such a thing as a cautious pioneer, *Depository Trust & Clearing Corporation* (DTCC) fills the bill. DTCC is the world's largest clearinghouse and depository, handling some $100 trillion in trading activity per year. To ensure the 24/7 continuity of its core business applications, in 1995 the company implemented two-way mirroring between IBM mainframes at two of its data centers. It implemented EMC's SRDF to handle mirroring across fiberoptic *Enterprise Systems Connection* (ESCON) connections. And recently, the company installed IBM's 2029 *Dense Wave Division Multiplexing* (DWDM) switches to optimize the use of dark fiber.

The company is just now completing the latest phase of its disaster-recovery strategy—the consolidation of 400 distributed servers and their direct-attached storage devices onto a SAN-based storage network infrastructure that also spans the two data centers. The servers handle file and print services, Lotus Notes, and internal application development, employing a combination of operating systems that include Microsoft Windows 2000, Windows NT 4.0, and Novell NetWare 4.11 (now being phased out).

"We wanted to build disaster recovery into applications and hardware from the ground up, which requires an enormous amount of infrastructure on the corporate level," says Michael Obiedzinski, DTCC's vice president of IT. "SANs were the only way to go."

Ensuring Redundancy

The infrastructure incorporates redundancy to address all the major potential points of failure. The company installed three Windows

2000 server clusters and dual SAN fabrics that extend across the two geographically dispersed data centers, which are linked via multi-strand fiber-optic DWDM connections. Twelve Brocade Silkworm 2800 Fibre Channel switches manage 10 TB of disk storage on EMC Clarion FC400 arrays.

"If a server fails, its resources can be picked up by a surviving server in the other city," explains Ed Alsberg, DTCC's IT operations director. "If a site or array fails, the surviving server is presented with a mirrored copy of the data. So we cover darn near any disaster scenario."

Computer Associates' ArcserveIT running on a Microsoft Windows 2000 server manages the storage subsystems and tape libraries. All backups to tape are written automatically to a Storagetek L700 tape silo at each data center. Each application has a primary server on one site and a secondary server on the other with two-way disk mirroring.

The SAN-based backup topology has been in production for 14 months and "works beautifully," Alsberg notes.

Callisma, the storage systems integrator that helped DTCC design its dual-SAN fabric, sees a growing number of its customers—75 percent, in fact—implementing similar configurations, says Jeffrey Birr, national director of storage networking for the integrator. "They've gotten more concerned about disaster recovery in the past few months," Birr declares. "They want to control things in-house rather than outsource, and they now have the technology to do it."

Seeking Safety and Soundness

While DTCC, with a 2-year head start, is in the vanguard of advanced backup implementations, Alsberg emphasizes that the company is usually very conservative when it comes to implementing innovative technologies. "Safety and soundness are top priorities for everything we do," he declares. "We move very slowly and cautiously."

For example, DTCC has no immediate plans to move its core business applications from IBM Z900 mainframes to open systems and SANs. With applications handling some $23 trillion in customer assets, "We don't make jumps to new technology lightly," Alsberg says.

The same goes for the company's Sun Solaris three-tiered Web platform, which provides customers with access to online financial services. "We're taking the approach that production customer-facing applications belong on our EMC Symmetrix/SRDF/ESCON storage infrastructure," Alsberg states. "ESCON and SRDF are tried and true technologies, while the SAN fabric is a new technology to us."

On the other hand, DTCC is highly confident that its dual-SAN fabric, which should be fully in place this quarter, will provide more than adequate backup and disaster recovery for Lotus Notes, file and print servers, and internal application development servers.

SANs are expected to provide several big advantages, such as enabling DTCC to administer four times as much storage without needing to expand its current core IT staff of four. Says Alsberg, "You get a single view of all the storage required for hundreds of servers. You always know how much storage you have available for new requests, and centralized backup is really important when you have hundreds of servers across two data centers."

Storage administrators use Brocade's Fabricwatch to trap events, such as a switch or port crashing, and send them to Hewlett-Packard's OpenView enterprise management system. For performance management, DTCC uses McData's SAN Navigator.

SANs also make it possible to allocate storage capacity among servers with much more flexibility, Alsberg says. "Today, you can only buy EMC storage in 36 and 72 GB capacity. If a server only needs 10 GB, the rest is wasted. On a SAN, you can give one server 10 GB and use the rest where it's needed."

According to Obiedzinski, moving from direct-attached to SAN-based storage will enable his firm to reduce wasted data space by 30 to 40 percent.

Boosting Backup Speed

DTCC also expects to realize a major boost in backup speeds by using SANs instead of 100 Mbps Ethernet. Right now, speeds are limited by the DLT 7000 tape drives, which are about to be replaced with Storagetek Super DLT drives. DTCC plans to further boost backup speeds by moving to server-less backup on Brocade's new 12,000 switches.

In order to implement server-less backup, however, DTCC must migrate the servers from their current Gigabit Ethernet connections to the SAN fabric. IT has held off so far because the servers need to communicate via *Transmission Control Protocol* (TCP)/*Internet Protocol* (IP) with the Arcserve IT host that manages backup. The clearinghouse is currently testing several Fibre Channel/IP technologies such as FCIP to ensure that whatever it deploys is ready for prime time, Alsberg comments.

Going for a High Level of Support

Testing is extremely important when dealing with Microsoft. "We want to qualify for a high level of Microsoft support, so everything in our Windows 2000 cluster configuration has to be on their Hardware Compatibility List," Alsberg explains. This means testing every relevant piece of server, storage, and SAN hardware and software right down to the exact model and revision.

"I understand that Microsoft needs to be absolutely sure that a particular configuration works," he says. "But that is not a trivial undertaking. In fact, Microsoft's original proposal was that we ship our entire configuration to Redmond so they could test it—$3 million worth of equipment." Fortunately, EMC volunteered to build a test bed at its Hopkinton, MA, facility, where it will test the entire configuration.

With the extended dual-SAN fabric and clustering infrastructure in place, DTCC, the cautious pioneer, is looking to the next major project: a third data center far enough away from the other two to be unaffected by regional disasters such as earthquakes.

"We're thinking of turning one of our existing centers into what EMC calls 'hop sites,' to do asynchronous SRDF over great distances," Alsberg declares. And what protocols will be used to make the hops? "We're looking at ESCON extension and FCIP, but we're leaning toward ESCON because it's proven technology," Alsberg says. "Again, if you're talking real money and business-critical applications, you don't want to go with the new kid on the block."

SAN-Bound A. B. Watley on Migration Path with Sun StorEdge T3 Arrays

Online trading requires a high-performance infrastructure that can process massive amounts of data in real time. However, the day-to-day operation of the system is not the only concern—storage issues are also important because of (1) the voluminous amount of data generated by online transactions and (2) the data generated for data-warehousing purposes demand a robust, highly available solution.

A.B. Watley Group, Inc., turned to Sun Microsystems when it was ready to streamline its infrastructure and storage setup. The company is the parent of A.B. Watley, Inc., a technology- and service-oriented brokerage firm that offers proprietary online trading systems such as UltimateTrader and WatleyTrader.

The firm began an across-the-board swap transition from Microsoft Windows NT to Sun in late 1998. According to Eric LeSatz, vice president of IT administration, the switch to Sun has translated to better availability and scalability and less maintenance over the long haul.

"We decided to revamp the entire trading system and go with a UNIX system," says LeSatz. "At the time, we were testing EMC, and we decided to bring Sun in to compare." High performance capacity, reliability, and scalability were crucial determinants in evaluating hardware that could meet the company's transactional and storage needs. The storage system takes each stock trade and inserts it into an Oracle database in real time. The system has to keep up with the demand or trades get dropped and data are no longer accurate.

A.B. Watley Group's trading infrastructure is built to handle the market fluctuations that occur daily. "At market open, we experience high volumes of quote and trade data," says LeSatz. "We also need to be prepared for major events such as interest-rate adjustments. The entire architecture was designed to support these peak-level data rates."

Selecting the Right Sun Hardware Solutions

The new system is currently storing approximately 2 TB of data on Sun StorEdge T3 arrays, with two-thirds of that devoted to data warehousing and one-third devoted to *Online Transaction Protocol* (OLTP). Each T3 array features a large built-in cache that minimizes *input/output* (I/O) response times so that requests and verifications of trades can be handled almost instantly. The company rewrote the client side of its proprietary applications in Java so that it could offer multiplatform support to clients based in Microsoft Windows, Macintosh, Linux, and Solaris.

The T3 is part of Sun's StorEdge family of network storage arrays that offer scalable high-availability storage solutions for all levels of business environments with systems ranging from 162 GB to 88 TB. The T3 array that A.B. Watley is using includes

- Two controller units in a partner-pair configuration, with two mirror-cache *redundant array of independent disks* (RAID) controllers linked together in path fail-over fashion
- Two enclosures
- Four hot-swappable redundant PCUs and UICs
- Two unit interconnect cables
- Eighteen hot-swappable RAID-ready dual-ported bidirectional Fibre Channel drives
- Two hot-swappable redundant RAID controller cards with 256 MB of mirrored cache
- External connections that enable the array to be connected to a data host and a management network

The T3 also features StorEdge Component Manager software, which provides a user-friendly graphic interface for easy monitoring and administration of the storage system, even remotely via Telnet and *File Transfer Protocol* (FTP). The unit supports IBM HACMP and Hewlett-Packard MC/Serviceguard HA clustering for high availability with A.B. Watley Group's mission-critical applications and storage.

Case Studies

Since LeSatz and the IT group installed its Sun T3 solution, the company has not had to revise its expectations between what it needed the servers to do and what they have been able to do. The company purchased *Small Computer System Interface* (SCSI)-based Sun A3500 arrays and later upgraded to the T3 arrays. The T3 arrays performed better for A.B. Watley's applications because of their Fibre Channel configuration and large cache.

The A3500 arrays include a variety of high-performance features needed to process and store daily trade traffic, including

- Fully redundant RAID configurations
- High-RAS hot-swappable power supplies, fan controllers, and disk drives
- RAID Manager software that supports RAID levels 0, 1, 1+0 and 3, 5
- Flexible configurations from 182 GB up to 5.6 TB rack-mounted systems
- Large storage capacity in a narrow footprint
- Support for up to 20 storage A3500 Fibre Channel arrays in one two-node cluster

Other Sun servers in use throughout the company's infrastructure include Sun Netra 1405 units working as communication servers for client access to the UltimateTrader trading system and Sun Enterprise 4500 transaction servers being used for trade request routing. LeSatz says that the T3 arrays are beneficial because their better performance and compatibility with Sun's SAN-focused Jiro technologies allow growth in a SAN environment.

The T3 array's reliability and availability features include

- Error checking and correction on disk drives
- Skip sectors and spare cylinders on disk drives
- Automatic sector reallocation on the RAID controller
- Link redundancy chip and 8- to 10-bit encoding on *Fibre Channel arbitrated-loops* (FC-ALs)
- Midplane and temperature sensors

The T3 array features built-in, hot-swappable, redundant *uninterruptible power source* (UPS) batteries to back up the cache in the event of a failure. The batteries power the controller unit and each of its nine disks, allowing contents in cache to be "destaged" to the disks if a power failure occurs. The T3 array's switched-loop architecture allows the A. B. Watley Group to scale in accordance to growth needs. In addition, its RAID controller card provides cache, RAID management, administration, diagnostics, and external interfaces. The controller card is also the data-processing and administrative brain of the array, in that it provides external interfaces and controls the system's back-end activities.

Planning for SAN

When the company moves to a SAN environment—which it hopes to do by the end of this year—LeSatz expects A.B. Watley to benefit from improved storage allocation. "Right now we tend to attach arrays directly to the host, and if we need more space than the array provides, we add a second unit," he says. "The SAN will allow us to pool our storage together and logically allocate partitions to whichever hosts require storage."

The firm has specific solid-state disk requirements. Because of this, it is more economical to put one larger disk on the SAN and allocate it between various machines instead of buying a smaller disk for each machine. There are other SAN benefits in the areas of backup and restore, but they are not yet top priorities. For now, LeSatz is waiting for more advanced SAN products to debut before beginning an implementation.

The switch change from Windows NT to Sun infrastructure was not as much a migration as a total rewrite. A.B. Watley replaced 200 Windows NT servers with attached storage with 10 Sun servers, making the management of storage, backup, and restore functions easier for the IT department. The Sun solution also takes up less room than the NT solution, comprising only three racks per each data center as opposed to the 22 racks that it took to house the NT system. The rack-to-user ratio changed considerably too. Sun's system supports a minimum of 25,000 users, whereas NT supported 5,000.

A.B. Watley Group keeps transactional data for 5 years, with each transaction running several hundred bytes. The other data—chart, time, and sales data—are kept for 30 days and then discarded. The company has data centers in Dallas and New York that complement each other and act as redundancy fail-overs.

So far the company is happy with the performance, availability, and reliability of the Sun architecture. As it plans its upcoming SAN migration, the firm is establishing a solid foundation based on compatible technologies and hardware that can scale to meet its future needs.

New SNIA Technology Center Opened for Interoperability Testing

On February 2, 2001, the *Storage Networking Industry Association* (SNIA) opened the doors to its new Technology Center in Colorado Springs, CO. It is the largest independent storage networking complex in the world. The function of the Technology Center is to provide a place where vendors and IT professionals can test storage interoperability.

"Testing the interoperability of large storage networking and server configurations is a main focus of the lab," says Thomas Conroy, director of the SNIA Technology Center.

The Technology Center enables industry players to prove and demonstrate interoperable configurations, architectures, education, and services for complete and proven storage networking solutions.

"We want to build a knowledge base of information on networked storage solutions to help propel the industry into a higher value-add position," says Conroy.

One example of how the Technology Center is going to deal with interoperability issues is events called *plug fests*. The purpose of these plug fests is to gather vendors in specific areas, such as IP storage, and have them use their various implementations of the technologies to see if the vendors' products interoperate.

For educational purposes, the Technology Center is going to conduct workshops to teach the vendor and user communities about various standards and interoperability. "These educational workgroups will be hands-on training in the labs and classrooms," says Conroy.

To test the architectures, the Technology Center will have permanent configurations of new technologies in place that will be used to test interoperability.

SNIA also plans to hold workgroup symposiums at the Technology Center four times a year. The workgroups will facilitate the study, development, and recommendation of storage standards for various standards bodies.

According to SNIA, early adopters have found that while storage networking technology promises to be beneficial by reducing maintenance costs and providing greater data and application availability, it can be very difficult to implement. This is where the laboratory comes in. Its vision is to "promote and accelerate the use of highly evolved, widely accepted storage network systems across the IT community and to solve real-world business problems."

For end users, the laboratory promises access to a multivendor networked test facility in which they can test new storage architectures in complex, realistic production environments. "There are main technology groups we want to explore in the Technology Center," says Donna Stever, chairperson of SNIA.

A sample of these technology groups includes disk resource management, Fibre Channel management, storage media library management, and Common Information Manager (CIM) demo. "The function of the Technology Center is to provide a place where people from these types of technology areas can come together and do work," says Stever.

SNIA is also planning to use the Technology Center as a place to validate standards it is working on, such as extended copy, Fiber Alliance MIB, and Host Bus Adapter API.

With more than 14,000 square feet, the state-of-the-art facility will include the following:

- 554 *local area network* (LAN) connections
- 186 Fibre Channel optical cable connections

- 800 × 20 Amp circuits, with a total available power in the laboratories and classroom at over 2,000 Amps

So far, SNIA has received some real positive feedback about the Technology Center. "Vendors are very excited about how far we have come with interoperability," says Stever.

Summary

Often case studies are a practical and effective tool for encouraging discussion about best practices and problem-solving strategies. It is suggested that you use them in the planning and implementation phases of your company's storage project.

GLOSSARY

Grateful acknowledgment is extended to the *Storage Networking Industry Association* (SNIA) for their permission to reprint this comprehensive glossary from their website.

Access The ability to use an information system resource.

Access control (ACL) Granting or withholding service, or access, to a resource by a requestor based on the identity of the administrator making the request.

Access control list A list that counts the rights of administrators (for example, users and groups of users) to access resources. Used in file systems to denote lists that are maintained by file systems and defines user and group permissions for file and directory accesses.

Access control mechanism A security safeguard designed to detect and deny unauthorized access, as well as permit authorized access, in an information system.

Access fairness A process by which nodes are guaranteed access to Fibre Channel arbitrated loops independently of other node activities.

Access method The way to access a transmission medium to transmit data.

Access path Host bus adapter, logical unit number, route through host-storage interconnect, controller, and logical unit used by a computer communicating with storage devices.

ACL Access control list.

ACS Automated cartridge system.

Active State of a Fibre Channel sequence initiator between start of transmission or reception frame of first data of a sequence and completion of a transmission or reception frame of last data in a sequence.

Active, active (components, controllers) Dual active components/controllers.

Active copper Type of Fibre Channel physical connection allowing up to 30 m of copper cable between adjacent devices.

Active, passive Hot standby components/controllers.

Active component System component requiring electrical power to operate; in storage subsystems, active components may include power supplies, storage devices, fans, and controllers.

Adapter Hardware device that converts timing and protocol of one bus or interface to another; adapters are implemented as specialized hardware on system boards as add-in cards to enable computer system processing hardware to access peripheral devices. An example of an adapter would be a host bus adapter.

Adapter card An adapter implemented as a printer circuit module

Adaptive array Disk array capable of changing its algorithm for virtual data address to physical data location mapping as the array is operating.

Address Fixed-length bit pattern that uniquely identifies a block of data stored on a disk or tape. Treated as a number, a mapping algorithm can be applied, concluding the physical location of a block of data. Small Computer System Interface (SCSI) byte value uniquely identifies a device connected to a SCSI bus for the purposes of communication. Applying this mapping algorithm to disk block addresses yields cylinder, head, and relative sector numbers where data may be found. Tape block addresses refer to relative positions of block data in single linear streams of blocks. Fixed-length bit pattern uniquely identifies a location (bit, byte, word) in computer memory.

Address identifier An address value used to identify source (S_ID) or destination (D_ID) of a frame. The FC-SW standard includes a table of special address identifier values and their meanings.

Address resolution Process of determining a Media Access Control (MAC) address.

Address Resolution Protocol (ARP) Protocol used to obtain mapping from a higher-layer address to a lower-layer address. There are four ARP messages for Internet Protocol (IP) running over Ethernet: ARP requests, replies, and reverses and ARP request and replies. ARP is used to refer to Ethernet Address Resolution Protocols. It is a protocol used by an IP networking layer to map IP addresses to lower-level hardware (MAC) addresses.

Addressing Algorithm in which areas of fixed disk, removable cartridge media, or computer system main memory are identified.

Administration host Computer that manages one or more storage subsystems (for example, filers, disk array subsystems, or tape subsystems).

Administrator Individual charged with installation, configuration, and management of a computer system, network, storage subsystem, database, or application.

Advanced Encryption Standard (AES) Cryptographic algorithm designated by the National Institute of Standards and Technology (NIST) as a replacement for Data Encryption Standard (DES).

Advanced intelligent tape (AIT) Tape device and media technology.

AES Advanced Encryption Standard.

Agent Program that performs one or more services (for example, gathering information from the Internet) acting as or for principal.

Aggregation Disk or tape data streams combined into single streams for higher performance. Two or more disks can be combined into single virtual disk for increased capacity. Two or more disks can be combined into a Redundant Array of Independent Disks (RAID) for high availability. Two or more input-output (I/O) requests for adjacently located data can be aggregated into single requests minimizing request processing overhead and rotational latency. Combining multiple similar and related objects or operations into a single object.

AH Authentication header.

AIT Advanced intelligent tape.

Algorithmic mapping Use of an algorithm to translate from one data addressing domain to another. If a volume is algorithmically mapped, the physical location of a block of data may be calculated from its virtual volume address using known characteristics of the volume (for example, stripe depth and number of member disks).

Alias An alternate name. Aliases are sometimes used for grouping purposes.

Alias address identifier One or more address identifiers that may be recognized by an N_Port along with its N_Port identifier. Alias address identifiers are used to form groups of N_Ports so that frames may be addressed to a group rather than individual N_Ports.

AL_PA Arbitrated loop physical address.

Alternate client restore Process of restoring files to a different client than the one from which they were backed up.

Alternate path restore The process of restoring files to a different directory than the one from which they were backed up.

Always on State of always having power applied (systems) or of being continually active (communication links); state of always being powered on and/or continually active. In a Fibre Channel, ESCON, or Gigabit Ethernet context, always on describes the state of an operational link. It is constantly transmitting either data frames, idles, or fills words. This can be contrasted with bursty transmissions and listening for a quiet line in Ethernet. For Fibre Channel management purposes, being always on allows link-level error detection on each transmitted word.

American National Standards Institute (ANSI) Coordinating organization for voluntary standards in the United States. Two ANSI working committees are aligned with storage-networking interests: X3T10 (responsible for SCSI I/O interface standards) and X3T11 (responsible for Fibre Channel interface standards).

ANSI American National Standards Institute.

ANSI T10 American National Standards Institute T10 Technical Committee, the standards organization responsible for SCSI standards for communication between computers, storage subsystems, and devices.

ANSI T11 American National Standards Institute T11 Technical Committee, the standards organization responsible for Fibre Channel moving electronic data into and out of computers, intelligent storage subsystems, and devices.

Glossary

ANSI X3T10 (T10) American National Standards Institute committee responsible for standards of accessing and controlling I/O devices; ANSI X3T10 is responsible for the SCSI family of standards.

ANSI X3T11 (T1) American National Standards Institute committee is responsible for standards for high-performance I/O interfaces such as Fibre Channel and High-Performance Parallel Interface (HPPI).

API Application programming interface.

Appliance Intelligent device programmed for performing single, well-defined functions (for example, file, Web, or print services). Appliances are considered to be capable of performing specialized functions at lower cost and with higher reliability than general-purpose servers are.

Application I/O request Application read request.

Application write request I/O requests made by storage clients, as distinguished from member I/O requests made by a storage subsystem's own control software. Storage Networking Industry Association (SNIA) publications do not generally distinguish between I/O requests made by the operating environment and those made by user applications.

Application programming interface (API) Interface used by an application program to request services. API is used to denote interfaces between applications and software components that comprise operating environments (for example, operating system, file system, volume manager, device drivers, and so on).

Application Response Measurement (ARM) Open Group technical standard defining function calls for transaction monitoring. The ARM standard is being advanced by both the Open Group and the Distributed Management Task Force.

Application-specific integrated circuit (ASIC) Integrated circuit designed for a particular application, such as interfacing to an SCSI bus.

Arbitrated loop Fibre Channel interconnect topology in which each port is connected to the next, forming a loop. Only one port in a Fibre Channel arbitrated loop can transmit data at any given time. Before transmitting data, a port in a Fibre Channel arbitrated loop must participate with all other ports in the loop in an arbitration to gain the right to transmit data. Arbitration logic is distributed among all of a loop's ports.

Arbitrated loop physical address (AL-PA) An 8-bit value used to identify a participating device in an arbitrated loop.

Arbitration Process in which a user of a shared resource negotiates with other users for the right to use the resource; a port connected to a shared bus must win arbitration before transmitting data on the bus.

Archive Collection of data taken to maintain a long-term record of a business or application state. Archives are used for auditing or analysis rather than for application recovery. Once files are archived, online copies of them are deleted and must be restored by explicit action.

Archiving Creating an archive.

ARM Application Response Measurement; common microprocessor architecture and the name of the company that created the architecture.

ARP Address Resolution Protocol.

Array Storage array (for example, disk array or tape array).

Array configuration Assignment of disks and operating parameters for a disk array. Disk array configuration designates the array's member disks or extents to which and order in which they are to use. Sets parameters of stripe depth, RAID model, cache allowance, spare disk assignments, and so on.

Association_Header Optional header used to associate Fibre Channel exchange with a process, system image, or multiexchange I/O operation on an end system; also used as part of exchange identifier management.

ASIC Application-specific integrated circuit.

Asymmetric cryptosystem Cryptographic algorithm in which different keys are used to block, encrypt, and decrypt a single mes-

Glossary

sage of stored information. One key remains a secret and private key; the other can be disclosed freely and is called a public key.

Asymmetric virtualization Out-of-band virtualization.

Asynchronous I/O request Request to perform an asynchronous I/O operation.

Asynchronous I/O operation I/O operation in which the initiator does not await completion before proceeding with other work. Asynchronous I/O operations enable an initiator to have multiple concurrent I/O operations in progress.

Asynchronous Transfer Mode (ATM) Connection-oriented data communication technology based on switching 53-byte fixed-length units of data called *cells*. Each cell is dynamically routed. ATM transmission rates are multiples of 51.840 Mbps. In the United States, a public communications service called Synchronous Optical Network (SONET) uses ATM at transmission rates of 155, 622, 2048, and 9196 Mbps. These are called OC-3, OC-12, OC-48, and OC-192, respectively. A similar service called Synchronous Digital Hierarchy (SDH) is offered in Europe. ATM is also used as a local area network (LAN) infrastructure, sometimes with different transmission rates and coding methods than are offered with SONET and SDH. More information is available from the ATM Forum.

ATM Asynchronous Transfer Mode.

Atomic operation Indivisible operation from an external perspective that occurs in its entirety or not at all. Database management systems implement the concept of business transactions and treat each business transaction as an atomic operation on the database. This means that all database updates that comprise transactions are performed or none of them are performed; it is never that some of them are performed and others are not. RAID arrays must implement atomic write operations to properly reproduce single-disk semantics from the perspective of their clients. Atomic operations are required to ensure that component failures do not corrupt stored data.

Attenuation Power dissipation between optical or electrical transmitters and receivers expressed in units of decibels (dB).

Audit trail Record of system activities in chronological format that enables reconstruction and examination of events and/or changes in an object sequentially. Audit trail may apply to information systems, communications systems, or sensitive material.

Authentication Process of determining what the requestor or provider of services represents. Security measure establishing validity of transmissions, messages, or originators or means of verifying individual authorization to receive information.

Authentication header (AH) Component of Internet Protocol Security (IPsec, standardized by the Internet Engineering Task Force) that permits specification of various authentication mechanisms designed to provide connectionless integrity, data origin authentication, and optional antireplay service.

Authorization Process of determining whether a requestor is allowed to receive a service or perform an operation. Access control is an example of authorization. Limiting of usage of information system resources to authorized users, programs, processes, or systems. Access control is a specific type of authorization.

Auto swap Automatic swap.

Automated cartridge system (ACS) Robot synonym.

Automatic backup Backup triggered by a schedule point or threshold reached rather than by human activity.

Automatic fail-over Fail-over that happens without human intervention.

Automatic swap Substitution of a replacement unit (RU) in a system for a defective one; substitution is performed by the system itself as it continues to perform normal functions. Automatic swaps are functional rather than physical substitutions; they do not require human intervention. Defective components must be replaced in a physical hot, warm, or cold swap operation.

Automatic switchover Automatic fail-over.

Availability Amount of time a system is available during time periods when it is expected to be available. Availability is measured as a percentage of an elapsed year. For example, 99.95 per-

Glossary

cent = 4.38 hours of downtime in a year ($0.0005 \times 365 \times 24 = 4.38$) for a system expected to be available all the time.

B_Port Port on a Fibre Channel bridge device.

Backing store Nonvolatile memory-backing store is used to contrast with a cache; it is a volatile random access memory used to speed up I/O operations. Data held in a volatile cache must be replicated or saved to a nonvolatile backing store so that they can survive system crashes or power failures. Data collection stored on (usually removable) nonvolatile storage media for purposes of recovery in case the original copy of data is lost or becomes inaccessible. To be useful for recovery, a backup must be made by copying the source data image when it is in a consistent state.

Backup client Computer system containing online data to be backed up.

Backup image Data collection that constitutes a backup copy of a given set of online data.

Backup manager An application program scheduling and managing backup operations.

Backup policy Rules for how and when backup should be performed; backup policies specify information such as what files or directories are to be backed up, schedule of backups, devices and media that are eligible to receive backup, how many copies should be made, and action to be perform if backup doesn't succeed.

Backup window Time available for performing a backup; backup windows are typically defined by operational necessity. If data were to be used from 8 A.M. until midnight, then the window between midnight and 8 A.M. is available for making backup copies. In some cases a backup window is an interval of time during which data and applications are unavailable.

Bandwidth Numerical difference between the upper and lower frequencies of a band of electromagnetic radiation; data transfer capacity.

Basic input-output system (BIOS) Small program residing in programmable, nonvolatile memory on a personal computer (PC) that is responsible for booting that computer and performing operating system-independent I/O operations. Standard BIOS interrupts are defined to allow access to a computer's disk, video, and other hardware components (for example, INT13 for disk access).

Baud Maximum rate of signal state changes per second on a communications circuit. If each signal state change corresponds to a code bit, then the baud rate and the bit rate are the same. It is possible for signal state changes to correspond to more than one code bit, so baud rate may be lower than code bit rate.

Bayonet Neil Councilman (BNC) connector Coaxial cable connector used in Ethernet applications. Specification for BNC connectors is contained in EIA/TIA 403-A and MIL-C-39012.

BB_Buffer Buffer associated with buffer-to-buffer flow control.

BB_Credit Buffer-to-buffer credit determines how many frames can be sent to a recipient when buffer-to-buffer flow control is in use.

Beginning running disparity Disparity present at a transmitter or receiver when an ordered set is initiated.

BER Bit error rate.

Berkeley RAID levels Disk array data protection and mapping techniques; there are six Berkeley RAID levels, referred to by names RAID level 1 through RAID level 6.

Best effort (class of service) A class of service that does not guarantee delivery of packets, frames, or datagrams; for the network, fabric, or interconnect to make all reasonable delivery efforts.

Big Endian Format for storage and transmission of binary data in which most significant bits are stored at the numerically lowest addresses or are transmitted first on a serial link.

BIOS Basic input-output system.

Bit error rate (BER) Likelihood that a transmitted bit will be erroneously received. BER is measured by counting the number of bits in error at the output of a receiver and dividing by the total number of bits in the transmission; typically expressed as a negative power of 10.

Glossary

Bit synchronization Process by which the receiver of a serial communication establishes its clocking used to locate code bits in a received data stream.

Black Designation applied to information systems in the context of security analysis to associated areas, circuits, components, and equipment where sensitive information is not processed.

Blind mating Ability of pairs of components to be connected without electrical or optical connection points being visible. Mechanical guides (for example, slots and rails) usually accomplish blind mating on components.

Block Unit where data are stored and retrieved on disk and tape devices; blocks are atomic units of data recognition (through preamble and block headers) and protection (through CRC or ECC). Units of application data from single information categories transferred within a single sequence.

Block addressing Algorithm uniquely identifying blocks of data stored on disk or tape media by number and then translation of those numbers into physical locations on the medium.

Block virtualization Applying virtualization (q.v.) to one or more block-based storage services for purposes of providing new aggregated, higher-level, richer, simpler, secure block service to clients; block virtualization functions can be nested. Disk drive, RAID system, or volume manager performs some form of block address to different block address mappings or aggregation.

BNC Bayonet Neil Councilman, a type of coaxial cable connector; the specifications for BNC connectors are defined in EIA/TIA 403-A and MIL-C-39012.

Boot, booting, bootstrapping Loading of code from disk or other storage device into a computer memory. Bootstrapping is an appropriate term because code load typically occurs in steps. Starting with a very simple program (BIOS) that initializes the computer's hardware and reads a sequence of data blocks from a fixed location on a predetermined disk into a fixed memory location. Data read is code for the next stages of bootstrapping—usually an operating system loader that completes the hardware setup, resulting in executing an operating system in memory.

Glossary

Bridge controller Storage controller that forms a bridge between two external I/O buses. Bridge controllers are used to connect single-ended SCSI disks to differential SCSI or Fibre Channel host I/O buses.

Broadcast Transmission of a message to all receivers (ports) connected to a communications facility at the same time. Broadcast can be contrasted with unicast (sending a message to a specific receiver) and multicast (sending a message to a select subset of receivers). In Fibre Channel contexts, a broadcast refers to sending a message to all the N_Ports connected to a fabric.

Buffer Solid-state memory device or programming construct used to hold data momentarily as they move along an I/O path or between software components. Buffers allow devices to communicate using links with faster or slower data transfer speeds, allows devices with different native processing speeds to intercommunicate, and allows software components to intercommunicate, share data, and coordinate activities.

Buffer-to-buffer flow control Flow control that occurs between two directly connected Fibre Channel ports (for example, an N_Port and an associated F_Port). A port indicates the number of frame buffers that can be sent to it (buffer credit) before sender is required to stop transmitting and wait for the receipt of a ready indication. Buffer-to-buffer flow control is used only when another NL-Local port logs onto an NL-Local port or when Fx ports log onto Nx ports.

Bypass circuit Circuit that removes a device from a data path (for example, Fibre Channel arbitrated loop) when valid signaling is lost.

Byte An 8-bit organizational unit for data; unit in which data are delivered to and by applications. Fibre Channel bytes are organized with the least significant bit denoted as bit 0 and the most significant bit as bit 7. (Most significant bit is shown on the left side in FC-PH documents.)

CA Certification authority.

Cable plant An installation's passive communications elements (for example, optical fiber, twisted pair, coaxial cable, connectors, or splices) between transmitters and receivers.

Glossary

Cache High-speed memory or storage device used to reduce effective time required to read data from or write data to a lower-speed memory or storage device. Read cache holds data in anticipation that they will be requested; Write cache holds data written by a client until they can be stored safely on more of a permanent storage medium such as a disk or tape.

Canister Enclosure for a single disk or tape usually designed to mount in shelves that supply power, cooling, and I/O bus services to devices. Canisters minimize radiofrequency emissions and simplify insertion and removal of devices in multidevice storage subsystems.

Carousel Medium that handles a robot in which the medium is stored and selected from a rotating wheel

Carrier Sense Multiple Access with Collision Detection (CSMA/CD) Physical layer data transmission protocol used in Ethernet and Fast Ethernet networks. Carrier sense refers to arbitration for shared links. Unlike always on physical protocols, Carrier Sense Protocols require a node wanting to transmit to wait for the absence of a carrier (indicating that another node is transmitting) on the link. Multiple access refers to the party-line nature of a link; a large number of nodes (up to 500 with Ethernet) share access to a single link. Collision detection refers to the possibility that two nodes will simultaneously sense an absence of the carrier and begin transmissions, interfering with one another. Nodes are required to detect this interference and then to cease transmission. With Ethernet, each node detecting a collision is required to wait for a random interval before attempting transmission again.

Cascading Process of connecting two or more Fibre Channel hubs or switches to increase the number of ports or to extend distances.

Catalog Stored list of backed-up files and directories with media identifiers of backup copies; backup managers use catalogs to determine what files must be backed up and what media must be mounted and read to perform restores; also persistent data structures used by file systems to keep track of the files they manage.

CDR Clock and data recovery.

Certification authority A public key infrastructure (PKI) authority and organization responsible for issuing and revoking user certificates to ensure compliance with PKI policies and procedures.

Changed block-changed block point-in-time copy Class of point-in-time copy implementations or resulting copies in which copy and source share storage for portions (blocks) of copy that are not modified (for example, from source if copy is writeable). Storage is physically copied as a consequence of modifications (for example, to source if copy is writeable). Changed block copy occupies only storage necessary to hold blocks of storage that have been changed since point-in-time where copy logically occurred.

Channel Electric circuits that sense or cause state changes in recording media and convert those state changes and electrical signals that can be interpreted as data bits. In I/O, the term *channel* has other meanings in other computer technology categories.

Character Byte, 10-bit information unit transmitted and received by FC-1; 8B/10B encoding provides mapping between 8 bits of data and a 10-bit transmission character. Transmission characters correspond to special codes, and not all 10-bit sequences represent valid transmission characters.

Character cell interface Command line interface.

Check data Data stored on member disks that can be used for regenerating user data that became inaccessible.

Checkpoint Recorded state of an application at an instant of time, including data, in-memory variables, program counter, and other contexts that would be required to resume application execution from the recorded state. Activity of a file system (such as high-performance file system [HPFS] or Andrews file system [AFS]) in which cached metadata (data about structures of file system) are periodically written to file system's permanent store, allowing file system consistency if an unexpected stops occur.

Chunk Strip.

Chunk size Strip depth and strip size.

C-H-S addressing Cylinder-head-sector addressing.

Glossary

CIFS Common Internet File System.

CIM Common Information Model.

Cipher Cryptographic system in which arbitrary symbols or groups of symbols represent units of plain text or units of plain text are rearranged or both.

Ciphertext Data security encrypted.

Circuit Communications circuit.

CKD (architecture) Count-key data disk architecture.

Class 1 Connection-oriented class of communications service in which the entire bandwidth of the link between two ports is dedicated to communication between ports and not used for other purposes; known as dedicated connection service. Class 1 service is not used widely.

Class 2 Connectionless Fibre Channel communications service that multiplexes frames from one or more N_Ports or NL_Ports; Class 2 frames are acknowledged by the receiver, and notification of delivery failure is provided. This class of service includes end-to-end flow control.

Class 3 Connectionless Fibre Channel communications service that multiplexes frames to or from one or more N_Ports or NL_Ports; Class 3 frames are datagrams—they are not explicitly acknowledged, and delivery is on a best-effort basis.

Class 4 Connection-oriented class of communications service in which a fraction of bandwidth of link between two ports is dedicated for communication between the ports. The remaining bandwidth may be used for other purposes; Class 4 service supports bounds on the maximum time to deliver a frame from sender to receiver known as *fractional service*; this service is not used widely.

Class 6 Connection-oriented class of communications service between two Fibre Channel ports that provides a dedicated unidirectional connection for reliable multicast known as *unidirectional dedicated connection service*.

Classified information Information determined to require protection against unauthorized disclosure and marked to indicate its classified status.

Class of service   Mechanism for managing traffic in a network by specifying message or packet priority, characteristic, and guarantee of transport layer of Fibre Channel Circuit. Fibre Channel classes of service include connection services (classes, guaranteed frame delivery with end-to-end flow control (class), packetized frame datagrams (class), and quality-of-service subchannels (for example, constant sub rate or constant latency) (class). Different classes of service may exist in the same fabric. Form and reliability of delivery in class 3 circuits may vary with topology; identification and grouping of data packets based on priority labels or other.

Cleartext   Data that are not encrypted.

CLI   Command line interface.

Client   Intelligent device or system requesting services from other intelligent devices, systems, or appliances; asymmetric relationships with a second party in which client initiates requests and server responds.

Client service request   Request issued by a client application to a well-known service.

Cluster   Collection of interconnected computers at high speeds for purposes of improving reliability, availability, serviceability, and/or performance through load balancing; clustered computers have access to common pools of storage and run software to coordinate component computer activities.

CMIP   Common Management Information Protocol.

Coaxial cable   Electrical transmission medium with two concentric conductors separated by a dielectric material in which spacings and material are arranged to give a specified electrical impedance.

Code balance   Number of 1 bits in a 10-bit transmitted data stream divided by 10 (for example, 1110100011 has a code balance of 6/10 = 60 percenmt).

Code bit   Bit or binary digit of encoded datum; sequences of code bits make up symbols, each corresponding to a data element (for example, words, bytes, or other); smallest time used by FC-0 for transmission on media.

Glossary

Code byte Byte of encoded datum called a symbol; code bytes are output encoding or encryption processes. Code bytes are referred to as *code words*.

Code violation Error condition that occurs when received transmission characters cannot be decoded into valid data bytes or special codes using validity checking rules specified by transmission codes.

Cold backup Offline backup.

Cold swap Substitution of replacement unit (RU) in a system for a defective one, where external power must be removed from the system in order to perform the substitution. Cold swap is a physical substitution and a functional substitution.

Comma character Either of 7-bit sequences 0011111 or 1100000 in an encoded stream; special character containing a comma.

Command line interface (CLI) Human interface to intelligent devices characterized by nondirective prompting and character string user input; many users believe to be more difficult to comprehend and use than graphic user interfaces (GUI).

Common Information Model (CIM) Object-oriented description of entities and relationships in business management environment maintained by Distributed Management Task Force. CIM is divided into a core model and common models; core model addresses high-level concepts (systems and devices) as well as fundamental relationships (dependencies). Common models describe specific problem domains (for example, computer system, network, user, or device management) and are subclasses of the core model or subclasses of one another.

Common Internet File System (CIFS) Network file system access protocol designed and implemented by Microsoft Corporation under Server Message Block Protocol and used by Windows clients to communicate file access requests to Windows servers. Implementations of CIFS protocol allow other clients and servers to use intercommunication and interoperation with Microsoft operating systems.

Common Management Information Protocol (CMIP) Network management protocol built on the Open Systems Interconnection (OSI) communication model. CMIP is more complete and therefore larger than Simple Network Management Protocol (SNMP).

Communication circuit Bidirectional path for message exchange within Fibre Channel fabric; networking, a specific logical or physical path between two points over communications occurs.

Communications security Protection of information being transmitted, particularly via telecommunications; the focus of communications security is message authenticity. Communications security includes cryptography, transmission security, emission security, and physical security.

Complex array Disk array whose control software protects and maps data according to complex algorithms; most common complex arrays are multilevel DISK ARRAYS that perform more than one level of data address mapping and adaptive arrays that are capable of changing data address mapping dynamically.

Compression Process of encoding data to reduce size; lossy compression (that is, compression using a technique in which a portion of the original information is lost) is acceptable for some forms of data (for example, digital images) in some applications, but in most information technology (IT) applications, lossless compression (that is, compression using a technique that preserves the entire content of the original data and from which original data can be reconstructed exactly) is required.

Computer security Measures and controls that ensure confidentiality, integrity, and availability of information system assets including hardware, software, firmware, and information being processed, stored, and communicated.

Concatenation Logical joining of two series of data, represented by the symbol | in data communications; two or more data items often are connected to provide unique names or references (for example, S_ID | X_ID); volume managers concatenate disk address spaces to present single, larger address spaces.

Concurrency Property of overlapping in time; refers to execution of I/O operations or I/O requests.

Glossary

Concurrent, concurrent copy Hybrid point-in-time copy mechanism in which each copy is initially a changed block copy (that is, shares unmodified storage with source) and over time becomes a split mirror copy (that is, does not share storage with source) without changing the point in time when the copy logically occurred independent of whether and where modifications to source or copy subsequently occur. Concurrent copy occupies the amount of storage required to hold changed blocks and grows to occupy as much storage as the copy source.

Concurrent operations Operations that overlap in time; concurrent I/O operations are central to use of independent access arrays in throughput-intensive applications.

Conditioning Processing of a signal for the purpose of making it conform more closely to an ideal; power conditioning is used to minimize voltage and frequency variations in an external power source; signal conditioning is used to reduce noise in logic or data signals.

Confidentiality Security encryption.

Configuration Installation or removal of hardware or software components required for system or subsystem functionality; assignment of operating parameters of a system, subsystem, or device. Disk array configuration (for example, includes array member disks or extents designation in addition to parameters such as stripe depth, RAID model, cache allowance, (and so on), collection of system hardware and software components and operating parameters.

Connection Dedicated connection.

Connection initiator N_Port initiates class 1 connection with destination N_Port through a connect request and receives valid response from destination N_Port to establish connection.

Connection recipient N_Port receives class 1 connect request from connection initiator and accepts connection request by transmitting a valid response.

Connectionless buffer Receive buffer used in a connectionless service capable of receiving connectionless frames.

Connectionless frame Frame used in connectionless service (that is, class 1 frames with SOF; class 2 and class 3 frames referred to individually or collectively).

Connectionless integrity service Security service that provides data integrity service for individual IP datagrams that detects modification of datagrams without regard to ordering of datagrams in a stream of datagrams.

Connectionless service Communication between two N_Ports or NL_Ports without a dedicated connection.

Console, enterprise management console Device for graphic or textual visual output from a computer system; for network and device management, an application that provides graphic and textual feedback regarding operation and status and may accept operator commands and input influencing operation and status.

Consolidation Process of accumulating data for a number of sequential write requests in a cache performing a smaller number of larger write requests to achieve efficient device utilization.

Continuously increasing relative offset Transmission control algorithm in which frames containing subblocks that comprise a block of information are transmitted strictly in the order of subblocks; continuously increasing relative offset simplifies reassembly and detection of lost frames relative to random relative offset.

Control software Body of software that provides common control and management for one or more disk arrays or tape arrays. Control software presents arrays of disks or tapes; it controls its operating environment as one or more virtual disks or tapes. Control software may execute in a disk controller, intelligent host bus adapter, or a host computer when it executes a disk controller or adapter. Control software is referred to as *firmware*.

Controller Control logic in a disk or tape that performs command decoding and execution, host data transfers, serialization and deserialization of data, error detection and correction, and overall management of device operations; control logic in storage subsystems that perform command transformation and routing, aggregation (RAID, mirroring, striping, or other), high-level error recovery, and performance optimization for multiple storage

Glossary

devices. Subclass of CIM_LogicalDevice, CIM_Controller represents a device having a single protocol stack whose primary purpose is to communicate with, control, and reset connected devices. (There are many subclasses of CIM_Controller addressing SCSI, PCI, USB, serial, parallel, and video controllers.)

Controller-based array, controller-based disk array Disk array control software that executes in a disk subsystem controller; member disks of a controller-based array are part of the same disk subsystem that includes the controller.

Controller cache Cache that resides in a controller with the primary purpose of improving disk or array I/O performance.

Copy on write Technique for maintaining a point-in-time copy or collection of data by copying only data that are modified after replicate initiation; original source data are used to satisfy read requests for the source data itself and the unmodified portion of point-in-time copy.

Copyback Replacement of properly functioning array member by another disk, including copying member's contents to replacing disk; copyback is used most often to create or restore particular physical configurations for arrays (for example, a particular arrangement of array members on device I/O buses) and is accomplished without a reduction of the array.

Count-key-data (CKD) Disk data organization model in which a disk is assumed to consist of a fixed number of tracks, each having maximum data capacity; multiple records of varying length may be written on each track of a CKD disk, and the usable capacity of each track depends on the number of records written to it. CKD)architecture derives its name from the record format, which consists of a field containing the number of bytes of data and a record address, an optional key field by which particular records can be easily recognized, and the data themselves. CKD is the Storage architecture used by IBM Corporation's System 390 series of mainframe computer systems.

Covert channel Unintended and/or unauthorized communications path that can be used to transfer information in a manner that violates security policy.

Credit Number of receive buffers allocated to transmitting N_Port, NL_Port, or F_Port; credit is the maximum number of outstanding frames that can be transmitted by an N_Port, NL_Port, or F_Port without causing a buffer overrun condition at the receiver.

CRC Cyclic redundancy check.

CRU Customer-replaceable unit.

Cryptanalysis Set of operations performed in converting encrypted information to plain text without initial knowledge of the algorithm and/or key employed in the encryption.

Cryptosystem A single means of encryption or decryption.

CSMA/CD Carrier Sense Multiple Access with Collision Detection.

Cumulative incremental backup Backup where data objects modified after the last full backup are copied; to restore data when cumulative incremental backups are in use, only the latest full backup and latest cumulative incremental backup are required.

Current running disparity Running disparity present at a transmitter when encoding of a valid data byte or special code is initiated or at a receiver when decoding of a transmission character is initiated.

Customer-replaceable unit (CRU) Unit or component of a system designed to be replaced by customers.

Cut through (switching) Switching technique allowing routing decisions to be made and acted on as soon as a destination address of a frame is received.

Cyclic redundancy check (CRC) Scheme for checking correctness of data that have been transmitted or stored and retrieved. CRC consists of a fixed number of bits computed as a function of the data to be protected and appended to the data. When the data are read or received, the function is recomputed, and the result is compared with that appended to the data. CRCs differ from error-correcting codes in that they can detect a wide range of errors but are not capable of correcting them.

Glossary

Cylinder-head-sector addressing A form of addressing data stored on a disk in which the cylinder, head/platter combination, and relative sector number on a track are specified.

D_ID Destination identifier.

Daemon A process that is always running on a computer system to service a particular set of requests. For example, in UNIX, lpd is a daemon that handles printing requests. Daemons are independent processes and not part of an application program. Application requests may be serviced by a daemon.

Data availability The amounts of time that data are accessible to applications during those time periods when they are expected to be available. Data availability is often measured as a percentage of an elapsed year. For example, 99.95 percent availability equates to 4.38 hours of unavailability in a year ($0.0005 \times 365 \times 24 = 4.38$) for a set of data that is expected to be available all the time.

Data byte A byte of user data as presented to a storage or communication facility. Data bytes are input to processes that encode for transmission or encrypt for privacy.

Data character Any transmission character associated by the transmission code with a valid data byte.

Data Encryption Standard (DES) A cryptographic data protection algorithm published by the National Institute of Standards and Technology (NIST) in Federal Information Processing Standard (FIPS) Publication No. 46.

Data frame A frame containing information meant for FC-4 (ULP) or the link application.

Data manager A computer program whose primary purpose is to present a convenient view of data to applications and to map that view to an internal representation on a system, subsystem, or device. File systems and database management systems are the most common forms of a data manager.

Data model A repository-specific representation of an Information model. A database representation of the CIM schemas is an example of a data model.

Data reliability Expressed as mean time to data loss (MTDL). The length of the statically expected continuous span of time over which data stored by a population of identical disk subsystems can be retrieved correctly.

Data stripe depth User data extent stripe depth.

Data striping A disk array data mapping technique in which fixed-length sequences of virtual disk data addresses are mapped to sequences of member disk addresses in a regular rotating pattern. Disk striping is commonly called RAID level 0 or RAID 0 because of its similarity to common RAID data mapping techniques. It includes no data protection, however, so strictly speaking, the appellation RAID is a misnomer.

Data transfer capacity The maximum rate at which data can be transmitted. Bandwidth is sometimes expressed in terms of signaling capacity (for example, SCSI) and sometimes in terms of data transmission capacity inclusive of protocol overhead (for example, Fibre Channel).

Data transfer-intensive (application) A characterization of applications. A data transfer-intensive application is an I/O-intensive application that makes large I/O requests. Data transfer-intensive applications' I/O requests are usually sequential.

Data transfer rate The amount of data per unit time actually moved across an I/O bus in the course of executing an I/O load. The data transfer capacity of an I/O subsystem is an upper bound on its data transfer rate for any I/O load. For disk subsystem I/O, data transfer rate is usually expressed in megabytes per second (millions of bytes per second, where 1 million = 10^6).

Database management system (DBMS) A set of computer programs with a user and/or programming interface that supports the definition of the format of a database and the creation of and access to its data. A DBMS removes the need for a user or program to manage low-level database storage; also provides security for and ensures the integrity of the data it contains. Types of DBMSs are relational (table-oriented) and object-oriented.

Glossary

Datagram A message sent between two communicating entities for which no explicit link-level acknowledgment is expected. Datagrams are said to be sent on a best-effort basis.

DBMS Database management system.

Decoding Validity checking of received transmission characters and generation of valid data bytes and special codes from those characters.

Decryption The operations performed in converting encrypted information to plain text with full knowledge of the algorithm and key(s) used to encrypt it. Decryption is the intended method for an authorized user to decrypt encrypted information.

Dedicated connection A communications circuit between two N_Ports maintained by a Fibre Channel fabric. The port resources used by a dedicated connection cannot be used for other purposes during the life of the dedicated connection.

Dedicated connection service Class 1 service.

Degraded mode Reduced mode. A mode of RAID array operation in which not all the array's member disks are functioning, but the array as a whole is able to respond to application read and write requests to its virtual disks.

Degaussing A procedure that reduces magnetic flux to virtually zero by applying a reverse magnetizing field—also called *demagnetizing*. Degaussing is used to ensure that no residual signal remains on magnetic media from which previously stored information could be recovered.

Delimiter An ordered set used to indicate a frame boundary.

DEN Directory-enabled network.

Denial of service (DoS) Result of any action or series of actions that prevents any part of an information system from functioning.

DES Data Encryption Standard.

Desktop Management Interface (DMI) A former name for the Distributed Management Task Force (DMTF).

Destination identifier (D_ID) An address contained in a Fibre Channel frame that identifies the destination of the frame.

Destination N_Port The N_Port to which a frame is addressed.

Device Storage device. CIM_LogicalDevice is an object that abstracts the configuration and operational aspects of hardware. Subclasses of CIM_LogicalDevice include low-level sensors, processors, and storage devices, and printer hardware.

Device bus Device I/O bus, an I/O bus used to connect storage devices to an HBA or intelligent controller. *Device I/O bus* is the preferred term.

Device channel A channel used to connect storage devices to a host I/O bus adapter or intelligent controller. The preferred term is *device I/O bus*.

Device fanout The ability of a storage controller to connect host computers to multiple storage devices using a single host I/O bus addresses. Device fanout allows computer systems to connect to substantially more storage devices than could be connected directly.

DHCP Dynamic Host Control Protocol.

Differential incremental backup A backup in which data objects modified since the last full backup or incremental backup are copied. To restore data when differential incremental backups are in use, the newest full backup and all differential backups newer than the newest full backup are required.

Differential (signaling) An SCSI electrical signaling technique in which each control and data signal is represented by a voltage differential between two signal lines. Differential signaling can be used over longer distances than the alternative single-ended signaling.

Differentiated Services (DiffServ) A protocol defined by the Internet Engineering Task Force (IETF) for managing network traffic based on the type of packet or message being transmitted; DiffServ rules define how a packet flows through a network based on a 6-bit field (the Differentiated Services code point) in the IP header. The DiffServ code point specifies the per-hop behavior (bandwidth, queuing, and forward/drop status) for the packet or message.

DiffServ Abbreviation for Differentiated Services.

Digest A computationally efficient function mapping of binary strings of arbitrary length to binary strings of some fixed length.

Digital linear tape (DLT) A family of tape devices and media technologies developed by Quantum Corporation.

Digital signature A cryptographic process used to ensure information authenticity, integrity, and nonrepudiation. Generally refers to assurances that can be externally verified by entities not in possession of the key used to sign the information. For example, a secure hash of the information encrypted with the originator's private key when an asymmetric cryptosystem is used. Some algorithms that are used in digital signatures cannot be used to encrypt data (for example, DSA).

Directory A mechanism for organizing information. A file or other persistent data structure in a file system that contains information about other files. Directories are usually organized hierarchically (that is, a directory may contain both information about files and other directories) and are used to organize collections of files for application or human convenience; Lightweight Directory Access Protocol (LDAP)-based repository consisting of class definitions and instances of those classes. An example of an enterprise-wide LDAP directory is Microsoft's Active Directory (AD) or Novell's NetWare Directory Service (NDS).

Directory-enabled network (DEN) An industry initiative, now part of the Distributed Management Task Force's (DMTF's) mission, to map the CIM schema to an LDAP directory; DEN's goals are to provide a consistent and standard data model to describe a network, its elements, and its policies/rules. Policies are defined to provide quality of service or to manage to a specified class of service.

Directory tree A collective term for a directory, all its files, and the directory trees of each of its subdirectories.

Discard policy An error-handling policy that allows an N_Port or NL_Port to discard data frames received following detection of a missing frame in a sequence.

Disconnection The process of removing a dedicated connection between two N_Ports.

Disk, disk drive A nonvolatile, randomly addressable, rewritable data storage device. This definition includes both rotating magnetic and optical disks and solid-state disks or nonvolatile electronic storage elements. It does not include specialized devices such as write-once-read-many (WORM) optical disks, nor does it include so-called RAM disks implemented using software to control a dedicated portion of a host computer's volatile random access memory.

Disk array A set of disks from one or more commonly accessible disk subsystems, combined with a body of control software. The control software presents the disks' storage capacity to hosts as one or more virtual disks. Control software is often called *firmware* or *microcode* when it runs in a disk controller. Control software that runs in a host computer is usually called a *volume manager*.

Disk array subsystem A disk subsystem that includes control software with the capability to organize its disks as disk arrays.

Disk block The unit where data are stored and retrieved on a fixed block architecture disk. Disk blocks are of fixed usable size (with the most common being 512 bytes) and are usually numbered consecutively. Disk blocks are also the unit of on-disk protection against errors; whatever mechanism a disk employs to protect against data errors (for example, ECC), it protects individual blocks of data.

Disk cache A cache that resides within a disk; a cache that resides in a controller or host whose primary purpose is to improve disk or array I/O performance.

Disk image backup A backup consisting of a copy of each of the blocks comprising a disk's usable storage area. A disk image backup incorporates no information about the objects contained on the disk and hence cannot be used for individual object restoration.

Disk shadowing Mirroring.

Disk striping Data striping.

Glossary

Disk subsystem A storage subsystem that supports only disk devices.

Disk scrubbing A function that reads all the user data and checks data blocks in a raid array and relocates them if media defects are found. Disk scrubbing can have a noticeably negative effect on application performance.

Disparity The difference between the number of 1s and the number of 0s in a transmission character.

Distributed Management Task Force (DMTF) Industry organization that develops management standards for computer system and enterprise environments. DMTF standards include WBEM, CIM, DMI, DEN, and ARM.

DLT Digital linear tape.

DMI Desktop Management Interface.

DMTF Distributed Management Task Force.

DNS Domain Name Service.

Document type definition (DTD) In eXtensible Markup Language (XML), a specification of the permissible tags or markup codes in a document and their meanings. Tags are delimited by the characters < and >. When a DTD is available for a document, a universal reader (program) can parse the document and display or print it.

Domain A shared user authorization database that contains users, groups, and security policies; a set of interconnected network elements and addresses that are administered together and that may communicate.

Domain controller A Windows NT or Windows 2000 server that contains a copy of a user account database. A Windows domain may contain both a primary and a backup domain controller.

Domain Name Service (DNS) A computer program that converts between IP addresses and symbolic names for nodes on a network in a standard way; most operating systems include a version of DNS.

DoS Denial of service.

Double buffering A technique often used to maximize data transfer rate by constantly keeping two I/O requests for consecutively addressed data outstanding. A software component begins a double-buffered I/O stream by making two I/O requests in rapid sequence. Thereafter, each time an I/O request completes, another is made immediately, leaving two outstanding. If a disk subsystem can process requests fast enough, double buffering allows data to be transferred at a disk or disk array's full volume transfer rate.

Drive letter A single letter of the alphabet by which applications and users identify a partition of physical or virtual disk to the Windows operating system. The number of letters in the alphabet limits the number of disks that can be referenced.

Driver Driver software, synonyms with *I/O driver*.

DSA Algorithm proposed by NIST in 1991 for use in digital signatures.

DTD Document type definition.

Dual active (components) A pair of components, such as the controllers in a failure-tolerant storage subsystem, that share a task or class of tasks when both are functioning normally. When one of the components fails, the other takes on the entire task. Dual active controllers are connected to the same set of storage devices and improve both I/O performance and failure tolerance compared with a single controller. Dual active components are also called *active-active components*.

Duplicate As a noun, a general term for a copy of a collection of data, including point-in-time copies; as a verb, the action of making a duplicate as just defined; any redundant component in a system.

Dynamic Host Control Protocol (DHCP) An Internet protocol that allows nodes to dynamically acquire (lease) network addresses for periods of time rather than having to preconfigure them. DHCP greatly simplifies the administration of large networks and networks in which nodes frequently join and depart.

Dynamic mapping A form of mapping in which the correspondence between addresses in two addresses' spaces can change over time.

Glossary

E_Port An expansion port on a Fibre Channel switch. E_Ports are used to link multiple Fibre Channel switches together into a fabric.

EBU European Broadcast Union.

ECC Error-correcting code.

EE_buffer A buffer associated with end-to-end flow control.

EE_credit A credit scheme used to manage end-to-end flow control during the exchange of frames between two communicating devices.

Electronic storage element Solid-state disk.

Embedded controller Embedded storage controller, an intelligent storage controller that mounts in a host computer's housing and attaches directly to a host's internal I/O bus. Embedded controllers obviate the need for HBAs and external host I/O buses. Embedded storage controllers differ from HBAs in that they provide functions beyond I/O bus protocol conversion (for example, RAID).

Encapsulating security payload (ESP) A component of IPsec that permits the specification of various confidentiality mechanisms.

Encoding Generation of transmission characters from valid data bytes and special codes.

Encryption The conversion of plain text to encrypted text with the intent that it only be accessible to authorized users who have the appropriate decryption key.

End of frame (EOF) A group of ordered sets that delineates the end of a frame.

End-to-end encryption Encryption of information at its origin and decryption at its intended destination without intermediate decryption.

End-to-end flow control Control of message flow between the two end parties to a communication on a network; flow control that occurs between two connected Fibre Channel N-Ports.

Enterprise resource management (ERM) Software that manages all aspects of an organization's assets, systems, services, and functions. The management of a set of resources in the wider perspective of an organization's entire business. Managing in an enterprise context requires that entities be named uniquely and locatable within the enterprise, that heterogeneity of platforms and services may be assumed, and that the dynamic nature of the environment is taken into account.

Enterprise Systems Connection (ESCON) A 200-Mbps serial I/O bus used on IBM Corporation's Enterprise System 9000 data center computers. Similar to Fibre Channel in many respects, ESCON is based on redundant switches to which computers and storage subsystems connect using serial optical connections.

Entry port-exit port A port in a media library through which media can be inserted or removed without exposing internal library components. Also called *exit port*.

EOF End of frame.

ERM Enterprise resource management.

Error-correcting code (ECC) A scheme for checking the correctness of data that have been stored and retrieved and correcting them if necessary; an ECC consists of a number of bits computed as a function of the data to be protected and appended to the data. When the data and ECC are read, the function is recomputed, the result is compared with the ECC appended to the data, and correction is performed if necessary. ECCs differ from CRCs in that the latter can detect errors but are not generally capable of correcting them.

ESCON Enterprise Systems Connection.

ESP Encapsulating Security Payload.

ESRM Enterprise storage resource management.

Ethernet The predominant LAN technology based on packetized transmissions between physical ports over a variety of electrical and optical media. Ethernet can transport any of several upper-layer protocols, the most popular of which is TCP/IP; Ethernet standards are maintained by the IEEE 802.3 Committee. The unqualified term *Ethernet* usually refers to 10-Mbps transmission on multipoint copper. *Fast Ethernet* is used to denote 100-Mbps transmission, also on multipoint copper facilities. Ethernet and Fast Ethernet both use CSMA/CD physical signaling. *Gigabit Ethernet* (GBE) transmits at 1250 MBd (1 Gb of data per second) using 8B/10B encoding with constant transmission detection.

Ethernet adapter   An adapter that connects an intelligent device to an Ethernet network, usually called an *Ethernet network interface card* or *Ethernet NIC*.

European Broadcast Union (EBU)   A European-based television (video) standardization group coordinated with the Society of Motion Picture and Television Engineers (SMPTE) and loosely affiliated with Fibre Channel Audio Video (FC-AV).

EVSN   External volume serial number, a set of one or more non-concurrent related sequences passing between a pair of Fibre Channel ports. An exchange encapsulates a "conversation" such as an SCSI task or an IP exchange. Exchanges may be bidirectional and may be short or long; the parties to an exchange are identified by an originator exchange identifier (OX_ID) and a responder exchange identifier (RX_ID).

Exchange identifier (X_ID)   A generic term denoting either an originator exchange identifier (OX_ID) or a responder exchange identifier (RX_ID).

Exchange status block   A data structure that contains the state of an exchange. An originator N_Port or NL_Port has an originator exchange status block and a responder N_Port or NL_Port has a responder exchange status block for each concurrently active exchange.

Exclusive connection   A Class 1 dedicated connection without intermix.

Exit port   A port in a media library where media can be inserted or removed without exposing internal library components.

Expansion card, expansion module   A collective term for optional adapters in the form of printer circuit modules that can be added to intelligent devices. Expansion cards include HBAs, NICs, and NVRAM, console, and other special-purpose adapters.

Expansion slot A mounting and internal bus attachment device within an intelligent device into which expansion cards are inserted.

Explicit addressing A form of addressing used with disks in which the data's address is explicitly specified in the access request.

Export To cause to appear or make available. Disk array control software exports virtual disks to its host environment. In file systems, a directory may be exported or made available for access by remote clients. To move objects, such as data, from within a system to a location outside the system, usually requiring a transformation during the move.

eXtensible Markup Language (XML) A universal format for structured documents and data on the World Wide Web (WWW); the World Wide Web (WWW) Consortium is responsible for the XML specification.

Extent A set of consecutively addressed fixed block architecture (FBA) disk blocks that is allocated to consecutive addresses of a single file; a set of consecutively located tracks on a CKD disk that is allocated to a single file; a set of consecutively addressed disk blocks that is part of a single virtual disk-to-member disk array mapping. A single disk may be organized into multiple extents of different sizes and may have multiple (possibly) nonadjacent extents that are part of the same virtual disk-to-member disk array mapping. This type of extent is sometimes called a *logical disk;* subclass or instance of the CIM_StorageExtent object. CIM model (removable and nonremovable) types of storage media.

External controller, external disk controller, external storage controller An intelligent storage controller that mounts outside its host computer's enclosure and attaches to hosts via external I/O buses. External storage controllers usually mount in the enclosure containing the disks they control.

External volume serial number (EVSN) A humanly readable volume serial number on a removable medium or cartridge.

Eye The region of an eye diagram that does not occur for correctly formed pulses. This is in the center of the eye diagram and distin-

Glossary

guishes presence of signal (region above the eye) from absence of signal (region below the eye).

Eye diagram A diagram used to specify optical or electrical pulse characteristics for transmitters. The horizontal axis represents normalized time from pulse start, and the vertical axis represents normalized amplitude.

Eye opening The time interval across the eye, measured at a 50 percent normalized eye amplitude, which is error free to the specified BER.

F_Port A port that is part of a Fibre Channel fabric. An F_Port on a Fibre Channel fabric connects to a node's N_Port. F_Ports are frame-routing ports and are insensitive to higher-level protocols; the link control facility within a fabric that attaches to an N_Port through a link. An N_Port uses a well-known address (hex'FFFFFE') to address the F_Port attached to it.

F_Port name A name identifier associated with an F_Port.

Fabric A Fibre Channel switch or two or more Fibre Channel switches interconnected in such a way that data can be physically transmitted between any two N_Ports on any of the switches. The switches comprising a Fibre Channel fabric are capable of routing frames using only the D_ID in an FC-2 frame header.

Fabric login (FLOGI) The process by which a Fibre Channel node establishes a logical connection to a fabric switch.

Fabric name A name identifier associated with a fabric.

Failback The restoration of a failed system component's share of a load to a replacement component. For example, when a failed controller in a redundant configuration is replaced, the devices that were originally controlled by the failed controller are usually failed back to the replacement controller to restore the I/O balance and to restore failure tolerance. Similarly, when a defective fan or power supply is replaced, its load, previously borne by a redundant component, can be failed back to the replacement part.

Failed over A mode of operation for failure-tolerant systems in which a component has failed and its function has been assumed by a redundant component. A system that protects against single failures operating in failed-over mode is not failure-tolerant because failure of the redundant component may render the system unable to function. Some systems (for example, clusters) are able to tolerate more than one failure; these remain failure-tolerant until no redundant component is available to protect against further failures.

Fail-over The automatic substitution of a functionally equivalent system component for a failed one; the term *fail-over* is applied most often to intelligent controllers connected to the same storage devices and host computers. If one of the controllers fails, fail-over occurs, and the survivor takes over its I/O load.

Failure tolerance The ability of a system to continue to perform its function (possibly at a reduced performance level) when one or more of its components has failed. Failure tolerance in disk subsystems is often achieved by including redundant instances of components whose failure would make the system inoperable coupled with facilities that allow the redundant components to assume the function of failed ones.

Fanout Device fanout.

Fast SCSI A form of SCSI that provides 10 megatransfers per second. Wide fast SCSI has a 16-bit data path and transfers 20 Mbps. Narrow fast SCSI transfers 10 Mbps.

Fault tolerance Failure tolerance.

FBA Fixed block architecture.

FC-PH The Fibre Channel physical standard, consisting of FC-0, FC-1, and FC-2.

FC-0 The Fibre Channel protocols level that encompasses the physical characteristics of the interface and data transmission media. Specified in FC-PH.

FC-1 The Fibre Channel protocols level that encompasses 8B/10B encoding and transmission protocol. Specified in FC-PH.

FC-2 The Fibre Channel protocol level that encompasses signaling protocol rules and the organization of data into frames, sequences, and exchanges. Specified in FC-PH.

Glossary

FC-3 The Fibre Channel protocols level that encompasses common services between FC-2 and FC-4. FC-3 contains no services in most implementations.

FC-4 The Fibre Channel protocol level that encompasses the mapping of upper-layer protocols (ULPs) such as IP and SCSI to lower-layer protocol (FC-0 through FC-). For example, the mapping of SCSI commands is a FC-4 ULP that defines the control interface between computers and storage.

FC-AE Fibre Channel Avionics Environment

FC-AL Fibre Channel arbitrated loop.

FC-AV Fibre Channel Audio Video.

FC-GS2 Fibre Channel Generic Services.

FC-SB, FC-SB2 Fibre Channel Single Byte (command set).

FC-SW, FC-SW2 Fibre Channel Switched (fabric interconnect).

FC-VI Fibre Channel Virtual Interface.

FCA Fibre Channel Association.

FCP Fibre Channel Protocol.

FCSI Fibre Channel Systems Initiative.

FDDI Fiber Distributed Data Interface.

FDDI adapter Adapter that connects an intelligent device to an FDDI network; both FDDI fiber adapters that connect to optical fiber FDDI networks and FDDI TP adapters that connect to twisted copper pair FDDI networks exist. Although network interface cards are usually referred to as NICs rather than as adapters, the term *FDDI adapter* is more common than FDDI NIC.

Federal Information Processing Standard (FIPS) Standards (and guidelines) produced by NIST for government-wide use in the specification and procurement of federal computer systems.

Federated Management Architecture Specification A specification from Sun Microsystems Computer Corporation that defines a set of Java APIs for heterogeneous storage resource and storage network management. This specification is a central technology of Jiro.

Fiber Distributed Data Interface An ANSI standard for a Token Ring metropolitan area networks (MANs) based on the use of optical fiber cable to transmit data at a rate of 100 Mbps. Both optical fiber and twisted copper pair variations of the FDDI physical standard exist. FDDI is a completely separate set of standards from Fibre Channel. The two are not directly interoperable.

Fiber A general term used to cover all transmission media specified in FC-PH.

Fibre Channel A set of standards for a serial I/O bus capable of transferring data between two ports at up to 100 Mbps, with standards proposals to go to higher speeds. Fibre Channel supports point-to-point, arbitrated loop, and switched topologies. Fibre Channel was developed completely through industry cooperation, unlike SCSI, which was developed by a vendor and submitted for standardization after the fact.

Fibre Channel Association (FCA) A former trade association incorporated in 1993 to promote Fibre Channel technology in the market. Separate FCA Europe and FCA Japan organizations also exist. In 1999, FCA merged with the Fibre Channel Loop Community (FCLC) to form the FCIA.

Fibre Channel Avionics Environment (FC-AE) A technical committee and industry group whose goal is to standardize Fibre Channel for avionics, defense, and other mobile applications.

Fibre Channel arbitrated loop (FC-AL) A form of Fibre Channel network in which up to 126 nodes are connected in a loop topology, with each node's L_Port transmitter connecting to the L_Port receiver of the node to its logical right. Nodes connected to a Fibre Channel arbitrated loop arbitrate for the single transmission that can occur on the loop at any instant using a Fibre Channel Arbitrated Loop Protocol that is different from Fibre Channel Switched and Point-to-Point Protocols. An arbitrated loop may be private (no fabric connection) or public (attached to a fabric by an FL_Port).

Fibre Channel Community, Fibre Channel Loop Community (FCLC) A former trade association incorporated in 1995 to promote Fibre Channel arbitrated loop technology for storage applications; the name was changed to Fibre Channel Community in 1997 to reflect changing goals and interests of the organization. In 1999, the FCLC merged with FCA to form the FCIA.

Fibre Channel Generic Services (FC-GS) An ANSI Standard that specifies several Fibre Channel services such as the name server, management server, time server, and others.

Fibre Channel Industry Association (FCIA) The industry association that resulted from the 1999 merger of the Fibre Channel Association and the Fibre Channel Community.

Fibre Channel name A name identifier that is unique in the context of Fibre Channel; essentially unused. Most Fibre Channel name identifiers are worldwide names that are unique across heterogeneous networks.

Fibre Channel Protocol (FCP) The serial SCSI command protocol used on Fibre Channel networks; FCP standardization is the responsibility of the X3T10 Committee.

Fibre Channel Service Protocol (FSP) An FC-4 protocol that defines all services independently of topology or fabric type.

Fibre Channel Single Byte (FC-SB) Command Set The industry standard command protocol for ESCON over Fibre Channel; a second version is known as FC-SB2.

Fibre Channel Switched (FC-SW) Fabric Interconnect The standard governing the form of Fibre Channel network in which nodes are connected to a fabric topology implemented by one or more switches. Each FC-SW node's N_Port connects to an F_Port on a switch. Pairs of nodes connected to an FC-SW network can communicate concurrently.

Fibre Channel Systems Initiative (FCSI) An industry association sponsored by Hewlett-Packard, IBM, and SUN with the goals of creating Fibre Channel profiles and promoting use of Fibre Channel for computer systems applications.

Fibre Channel Virtual Interface (FC-VI) A proposed standard for application-level distributed interprocess communication based on Intel Corporation's V1.0 Virtual Interface (VI) architecture; formerly known as VIA.

Fibre Connect (FICON) IBM Corporation's implementation of ESCON over Fibre Channel, standardized as Fibre Channel Single Byte Command Set.

FICON Fibre Connect.

Field-replaceable unit (FRU) A unit or component of a system that is designed to be replaced in the field, that is, without returning the system to a factory or repair depot. Field-replaceable units may either be customer-replaceable or their replacement may require trained service personnel.

File An abstract data object made up of (1) an ordered sequence of data bytes stored on a disk or tape, (2) a symbolic name by which the object can be uniquely identified, and (3) a set of properties, such as ownership and access permissions, that allows the object to be managed by a File System or backup manager. Unlike the permanent address spaces of storage media, files may be created and deleted and, in most file systems, may expand or contract in size during their lifetimes.

File server A computer whose primary purpose is to serve files to clients. A file server may be a general-purpose computer that is capable of hosting additional applications or a special-purpose computer capable only of serving files.

File system A software component that imposes structure on the address space of one or more physical or virtual disks so that applications may deal more conveniently with abstract named data objects of variable size (files). File systems are often supplied as operating system components but are implemented and marketed as independent software components.

File system virtualization The act of aggregating multiple file systems into one large virtual file system. Users access data objects through the virtual file system; they are unaware of the underlying; the act of providing additional new or different functionality, for example, a different file access protocol, on top of one or more existing file systems.

File virtualization The use of virtualization to present several underlying file or directory objects as one single composite file; the use of virtualization to provide HSM-like properties in a storage

Glossary

system; the use of virtualization to present an integrated file interface when file data and metadata are managed separately in the storage system.

Filer An intelligent network node whose hardware and software are designed to provide file services to client computers. Filers are preprogrammed by their vendors to provide file services and are not normally user-programmable.

Firmware Low-level software for booting and operating an intelligent device; firmware generally resides in read-only memory (ROM) on a device.

Fill byte Fill word, a transmission word that is an idle or an ARBx primitive signal. Fill words are transmitted between frames, primitive signals, and primitive sequences to keep a Fibre Channel network active.

FIPS Federal Information Processing Standard.

Fixed block architecture (FBA) A model of disks in which storage space is organized as linear, dense address spaces of blocks of a fixed size; FBA is the disk model on which SCSI is predicated.

FL_Port Port that is part of a Fibre Channel fabric; FL_Port on a Fibre Channel fabric connects to an arbitrated loop. Nodes on the loop use NL_Ports to connect to the loop. NL_Ports give nodes on a loop access to nodes on the fabric to which the loop's FL_Port is attached.

FLOGI Fabric login.

Formatting The preparation of a disk for use by writing required information on the medium. Disk controllers format disks by writing block header and trailer information for every block on the disk. Host software components such as volume managers and file systems format disks by writing the initial structural information required for the volume or file system to be populated with data and managed.

Frame An ordered vector of words that is the basic unit of data transmission in a Fibre Channel network. A Fibre Channel frame consists of a start-of-frame (SOF) word (40 bits), a frame header (8 words or 320 bits), data (0 to 524 words or 0 to 2192 10-bit encoded bytes, a CRC (1 word or 40 bits), and an end-of-frame (EOF) word (40 bits).

Frame content The information contained in a frame between its SOF and EOF delimiters, excluding the delimiters.

FRU Field-replaceable unit.

FSP Fibre Channel Service Protocol.

Full backup A backup in which all of a defined set of data objects are copied, regardless of whether they have been modified since the last backup. A full backup is the basis from which incremental backups are taken.

Full duplex Concurrent transmission and reception of data on a single link.

Full volume transfer rate The average rate at which a single disk transfers a large amount of data (for example, more than one cylinder) in response to one I/O request. The full-volume data transfer rate accounts for any delays (for example, due to inter-sector gaps, intertrack switching time, and seeks between adjacent cylinders) that may occur during the course of a large data transfer. Full volume transfer rate may differ depending on whether data are being read or written. If this is true, it is appropriate to speak of full-volume read rate or full-volume write rate. Also known as *spiral data transfer rate*.

G_Port A port on a Fibre Channel switch that can function either as an F_Port or as an E_Port. The functionality of a G_Port is determined during port login. G_Port functions as an F_Port when connected to an N_Port and as an E_Port when connected to an E_Port.

GbE Gigabit Ethernet.

Gb, Gbit, gigabit Shorthand for 1,000,000,000 (10^9) bits. Storage Networking Industry Association publications typically use the term Gbit to refer to 10^9 bits rather than 1,073,741,824 (2^{30}) bits; for Fibre Channel, 1,062,500,000 bits per second.

GB, Gbyte, gigabyte Shorthand for 1,000,000,000 (10^9) bytes. The Storage Networking Industry Association uses Gbyte to refer to

Glossary

10^9 bytes, as is common in I/O-related applications, rather than the 1,073,741,824 (2^{30}) convention sometimes used in describing computer system RAM.

GBIC   Gigabyte interface converter.

Geometry   The mathematical description of the layout of blocks on a disk; the primary aspects of a disk's geometry are the number of recording bands and the number of tracks and blocks per track in each, the number of data tracks per cylinder, and the number and layout of spare blocks reserved to compensate for medium defects.

Gigabaud link module (GLM)   A transceiver that converts between electrical signals used by HBSs (and similar Fibre Channel devices) and either electrical or optical signals suitable for transmission; GLMs allow designers to design one type of device and adapt it for either copper or optical applications. GLMs are used less often than gigabit interface converters because they cannot be hot swapped.

Gigabit   Gb.

Gigabit Ethernet (GbE)   A group of Ethernet standards in which data are transmitted at 1 Gbps. GbEcarries data at 1250 MBd using an adaptation of the Fibre Channel physical layer (8B/10B encoding); GBE standards are handled by IEEE 802.3z.

Gigabit interface converter (GBIC)   Transceiver that converts between electrical signals used by HBAs (and similar Fibre Channel and Ethernet devices) and either electrical or optical signals suitable for transmission; GBICs allow designers to design one type of device and adapt it for either copper or optical applications. Unlike gigabaud link modules (GLMs), GBICs can be hot swapped and are therefore gradually supplanting the former type of transceiver.

Gigabyte   GB.

Gigabyte System Network (GSN)   A common name for the HIPPI 6400 Standard for 800-Mbps links; a network of devices that implement the HIPPI 6400 Standard.

GLM   Gigabaud link module.

Graphic user interface (GUI) A form of user interface to intelligent devices characterized by pictorial displays and highly structured forms oriented input. Valued for perceived ease of use compared with a character cell interface.

Group A collection of computer user identifiers used as a convenience in assigning resource access rights or operational privileges.

GSN Gigabyte System Network.

GUI Graphic user interface.

Hacker An unauthorized user who attempts to gain and/or succeeds in gaining access to an information system.

Hard zone Zone consisting of zone members that are permitted to communicate with one another via the fabric. Hard zones are enforced by fabric switches, which prohibit communication among members not in the same zone. Well-known addresses are implicitly included in every zone.

HBA Host bus adapter.

Hierarchical storage management (HSM) The automated migration of data objects among storage devices usually based on inactivity; HSM is based on the concept of a cost-performance storage hierarchy. By accepting lower access performance (higher access times), one can store objects less expensively. By automatically moving less frequently accessed objects to lower levels in the hierarchy, higher-cost storage is freed for more active objects and a better overall cost: Performance ratio is achieved.

High availability The ability of a system to perform its function continuously (without interruption) for a significantly longer period of time than the reliabilities of its individual components would suggest. High availability is achieved most often through failure tolerance. High availability is not an easily quantifiable term. Both the bounds of a system that is called highly available and the degree to which its availability is extraordinary must be clearly understood on a case-by-case basis.

High-Performance Parallel Interface (HIPPI) An ANSI Standard for an 800-Mbps I/O interfaces used primarily in supercom-

Glossary

puter networks; the subsequent 6400-Mbps I/O interface standard, HIPPI-6400, is referred to more commonly as the Gigabyte System Network (GSN) standard.

High-speed serial direct connect (DSSDC) A form factor that allows quick connect/disconnect for Fibre Channel copper interfaces.

HIPPI High-Performance Parallel Interface.

Host A host computer.

Host adapter Host bus adapter.

Host-based array Host-based disk array volume. A disk array whose control software executes in one or more host computers rather than in a disk controller. The member disks of a host-based array may be part of different disk subsystems.

Host-based virtualization Virtualization implemented in a host computer rather than in a storage subsystem or storage appliance. Virtualization can be implemented either in host computers, in storage subsystems or storage appliances, or in specific virtualization appliances in the storage interconnect fabric.

Host bus Host I/O bus.

Host bus adapter (HBA) An I/O adapter that connects a host I/O bus to a computer's memory system; *host bus adapter* is the preferred term in SCSI contexts. *Adapter* and *NIC* are the preferred terms in Fibre Channel contexts. The term *NIC* is used in networking contexts such as Ethernet and Token Ring.

Host cache A cache that resides within a host computer whose primary purpose is to improve disk or array I/O performance. Host cache may be associated with a file system or database, in which case the data items stored in the cache are file or database entities. Alternatively, host cache may be associated with the device driver stack, in which case the cached data items are sequences of disk blocks.

Host computer Any computer system to which disks, disk subsystems, or file servers are attached and accessible for data storage and I/O. Mainframes, servers, workstations, and personal computers, as well as multiprocessors and clustered computer complexes, are all referred to as host computers in SNIA publications.

Host environment A storage subsystem's host computer or computers, inclusive of operating system and other required software instance(s). The term *host environment* is used in preference to host computer to emphasize that multiple host computers are being discussed or to emphasize the importance of the operating system or other software in the discussion.

Host I/O bus An I/O bus used to connect a host computer's HBA to storage subsystems or storage devices.

Hot backup Online backup.

Hot disk A disk whose capacity to execute I/O requests is saturated by the aggregate I/O load directed to it from one or more applications.

Hot file A frequently accessed file. Hot files are generally the root cause of hot disks, although this is not always the case. A hot disk also can be caused by operating environment I/O, such as paging or swapping.

Hot spare (disk) A disk being used as a hot standby component.

Hot standby (component, controller) A redundant component in a failure-tolerant storage subsystem that is powered and ready to operate but which does not operate as long as a companion primary component is functioning. Hot standby components increase storage subsystem availability by allowing systems to continue to function when a component such as a controller fails. When the term *hot standby* is used to denote a disk, it specifically means a disk that is spinning and ready to be written to, for example, as the target of a rebuilding operation.

Hot swap The substitution of a replacement unit (RU) in a system for a defective unit, where the substitution can be performed while the system is performing its normal functioning normally. Hot swaps are physical operations typically performed by humans.

Hot-swap adapter An adapter that can be hot swapped into or out of an intelligent device. Some adapters that are called hot-swap adapters more properly should be termed *warm-swap*

Glossary

adapters because the function they perform is interrupted while the substitution occurs.

HSM   Hierarchical storage management.

HSSDC   High-speed serial direct connect.

HTML   HyperText Markup Language.

HTTP   HyperText Transfer Protocol.

Hub   A communications infrastructure device to which nodes on a multipoint bus or loop are physically connected. Commonly used in Ethernet and Fibre Channel networks to improve the manageability of physical cables. Hubs maintain the logical loop topology of the network of which they are a part while creating a "hub and spoke" physical star layout. Unlike switches, hubs do not aggregate bandwidth. Hubs typically support the addition or removal of nodes from the bus while it is operating.

Hub port   A port on a Fibre Channel hub whose function is to pass data transmitted on the physical loop to the next port on the hub. Hub ports include loop-healing port bypass functions. Some hubs have additional management functionality. There is no definition of a hub port in any Fibre Channel standard.

Hunt group   A set has associated N_Ports in a single node attached to the same fabric. A hunt group is assigned a special alias address identifier that enables a switch to route any frames containing the identifier to be routed to any available N_Port in the group. FC-PH does not presently specify how a hunt group can be realized.

HyperText Markup Language (HTML)   The set of tags or "markup" codes that describe how a document is displayed by a web browser. Tags are delimited by the characters < and >. For example, the markup code <p> indicates that a new paragraph is beginning, whereas </p> indicates that the current paragraph is ending.

HyperText Transfer Protocol (HTTP)   An application-level protocol, usually run over TCP/IP, that enables the exchange of files via the World Wide Web (WWW).

ICMP   Internet Control Message Protocol.

Idempotency A property of operations on data; an idempotent operation is one that has the same result no matter how many times it is performed on the same data. Writing a block of data to a disk is an idempotent operation, whereas writing a block of data to a tape is not because writing a block of data twice to the same tape results in two adjacent copies of the block.

Idle, idle word An ordered set of four transmission characters normally transmitted between frames to indicate that a Fibre Channel network is idle.

IETF Internet Engineering Task Force.

Ignored (field) A field that is not interpreted by its receiver.

IKE Internet Key Exchange.

Implicit addressing A form of addressing usually used with tapes in which the data's address is inferred from the form of the access request. Tape requests do not include an explicit block number but instead specify the next or previous block from the current tape position, from which the block number must be inferred by device firmware.

In-band virtualization Virtualization functions or services that are in the data path. In a system that implements in-band virtualization, virtualization services such as address mapping are performed by the same functional components used to read or write data.

Incremental backup A collective term for cumulative incremental backups and differential incremental backups. Any backup in which only data objects modified since the time of some previous backup are copied.

Independent access array A disk arrays whose data mapping is such that different member disks can execute multiple application I/O requests concurrently.

Infinite buffer A term indicating that at the FC-2 level, the amount of buffering available at the sequence recipient is assumed to be unlimited. Buffer overrun must be prevented by each ULP by choosing an appropriate amount of buffering per sequence based on its maximum transfer unit size.

Glossary

Information category A frame header field indicating the category to which the frame payload belongs (for example, solicited data, unsolicited data, solicited control, and unsolicited control).

Information model A repository-independent definition of entities (that is, objects) and the relationships and interactions between these entities. For example, the CIM schemas are an example of an information model. An information model differs from a data model that is repository-specific.

Information system The entire infrastructure, organization, personnel, and components for the collection, processing, storage, transmission, display, dissemination, and disposition of information.

Information technology (IT) All aspects of information creation, access, use, storage, transport, and management. The term *information technology* addresses all aspects of computer and storage systems, networks, users, and software in an enterprise.

Information unit A related collection of data specified by FC-4 to be transferred as a single FC-2 sequence.

Infrastructure-based virtualization Virtualization implemented in the storage fabric, in separate devices designed for the purpose, or in network devices. Examples are separate devices or additional functions in existing devices that aggregate multiple individual file system appliances or block storage subsystems into one such virtual service, functions providing transparent block or file system mirroring functions, or functions that provide new security or management services.

Inherent cost The cost of a system expressed in terms of the number and type of components it contains. The concept of inherent cost allows technology-based comparisons of disk subsystem alternatives by expressing cost in terms of number of disks, ports, modules, fans, power supplies, cabinets, and so on. Because it is inexpensively reproducible, software is generally assumed to have negligible inherent cost.

Initial relative offset The relative offset of the block or subblock transmitted by the first frame in a sequence. The initial relative offset is specified by a ULP and need not be zero.

Initialization Startup and initial configuration of a device, system, piece of software, or network; for FC-1, the period beginning with power-on and continuing until the transmitter and receiver at that level become operational.

Initiator The system component that originates an I/O command over an I/O bus or network. I/O adapters, NICs, and intelligent controller device I/O bus control ASICs are typical initiators.

I_Node A persistent data structure in a UNIX or UNIX-like file system that describes the location of some or all of the disk blocks allocated to the file.

Instantiation The creation of an instance of a class- or object-oriented abstraction.

Intelligent controller A storage controller that includes a processor or sequencer programmed to enable it to handle a substantial portion of I/O request processing autonomously.

Intelligent device A computer, storage controller, storage device, or appliance.

Intelligent peripheral interface (IPI) A high-performance standards-based I/O interconnect.

Intercabinet A specification for Fibre Channel copper cabling that allows up to 30 m of distance between two enclosures that contain devices with Fibre Channel ports.

Interconnect A physical facility by which system elements are connected together and through which they can communicate with each other. I/O buses and networks are both interconnects.

Interface connector An optical or electrical connector that connects the media to the Fibre Channel transmitter or receiver. An interface connector consists of both a receptacle and a plug.

Intermix A Fibre Channel class of service that provides a full-bandwidth dedicated class 1 connection but allows connectionless class 2 and class 3 traffic to share the link during intervals when bandwidth is unused.

International Standards Organization (ISO) The International Standards body; ISO-published standards have the status of international treaties.

Glossary

Internet Control Message Protocol (ICMP) Control protocol strongly related to IP and TCP and used to convey a variety of control and error indications.

Internet Engineering Task Force (IETF) A large open international community of network designers, operators, vendors, and researchers concerned with evolution and smooth operation of the Internet and responsible for producing RFCs. The standards body responsible for Internet standards, including SNMP, TCP/IP, and policy for QoS; the IETF has a web site at www.ietf.org.

Internet Key Exchange (IKE) A protocol used to obtain authenticated keying material. Standardized by the Internet Engineering Task Force and described in RFC 2409.

Internet Protocol (IP) A protocol that provides connectionless best-effort delivery of datagrams across heterogeneous physical networks.

Interrupt Hardware or software signals that cause a computer to stop executing its instruction stream and switch to another stream. Application or other programs triggers software interrupts. Hardware interrupts are caused by external events to notify software so that it can deal with the events. The ticking of a clock, completion or reception of a transmission on an I/O bus or network, application attempts to execute invalid instructions or reference data for which they do not have access rights, and failure of some aspect of the computer hardware itself are all common causes of hardware interrupts.

Interrupt switch A human-activated switch present on some intelligent devices that is used to generate interrupts. Usually used for debugging purposes.

Intracabinet A Fibre Channel specification for copper cabling that allows up to 13 m of total cable length within a single enclosure that may contain multiple devices.

I/O Input-output; process of moving data between a computer system's main memory and an external device or interface such as a storage device, display, printer, or network connected to other computer systems. I/O is a collective term for reading or moving data into a computer system's memory and writing or moving data from a computer system's memory to another location.

I/O adapter Adapter that converts between the timing and protocol requirements of an intelligent device's memory bus and those of an I/O bus or network. In the context of storage subsystems, I/O adapters are contrasted with embedded storage controllers, which not only adapt between buses but also perform transformations such as device fan-out, data caching, RAID, and HBA.

I/O bus Any path used to transfer data and control information between components of an I/O subsystem. An I/O bus consists of wiring (either cable or backplane) connectors and all associated electrical drivers, receivers, transducers, and other required electronic components. I/O buses are typically optimized for the transfer of data and tend to support more restricted configurations than networks. In this book, an I/O bus that connects a host computer's HBA to intelligent storage controllers or devices is called a *host I/O bus*. An I/O bus that connects storage controllers or host I/O bus adapters to devices is called a *device I/O bus*.

I/O bottleneck Any resource in the I/O path (for example, device driver, HBA, I/O bus, intelligent controller, or disk) whose performance limits the performance of a storage subsystem as a whole.

I/O driver A host computer software component (usually part of an operating system) whose function is to control the operation of peripheral controllers or adapters attached to the host computer. I/O drivers manage communication and data transfer between applications and I/O devices using HBAs as agents. In some cases, drivers participate in data transfer, although this is rare with disk and tape drivers, since most HBAs and controllers contain specialized hardware to perform data transfers.

I/O intensity Characterization of applications; an I/O-intensive application is one whose performance depends strongly on the performance of the I/O subsystem that provides its I/O services. I/O-intensive applications may be data transfer-intensive or I/O request-intensive.

Glossary

I/O load A sequence of I/O requests made to an I/O subsystem. The requests that comprise an I/O load include both user I/O and host overhead I/O, such as swapping, paging, and file system activity.

I/O load balancing Load balancing.

I/O operation Read, write, or control function performed to, from, or within a computer system. For example, I/O operations are requested by control software in order to satisfy application I/O requests made to virtual disks.

I/O request A request by an application to read or write a specified amount of data. In the context of real and virtual disks, I/O requests specify the transfer of a number of blocks of data between consecutive disk block addresses and contiguous memory locations.

I/O subsystem A collective term for the set of devices and software components that operate together to provide data transfer services. A storage subsystem is one type of I/O subsystem.

IP Internet Protocol.

IPI Intelligent Peripheral Interface.

IPsec IP Security.

IP Security (IPSec) A suite of cryptographic algorithms, protocols, and procedures used to provide communication security for IP-based communications. Standardized by the Internet Engineering Task Force.

ISO International Standards Organization.

IT Information technology.

Java An object-oriented computer programming language that is similar to but simpler than C++. Java was created by Sun Microsystems Computer Corporation.

JBOD Just a bunch of disks. Originally used to mean a collection of disks without the coordinated control provided by control software; today, the term *JBOD* most often refers to a cabinet of disks, whether or not RAID functionality is present.

Jini An architecture and supporting services for publishing and discovering devices and services on a network. Jini was created by Sun Microsystems Computer Corporation.

Jiro A Sun Microsystems Computer Corporation initiative developed using the Java community process. Jiro's goal is to enable the management of heterogeneous storage networks. The core technology in Jiro is defined in the Federated Management Architecture Specification.

Jitter Deviation in timing that a bit stream encounters as it traverses a physical medium.

K28.5 A special 10-bit character used to indicate the beginning of a Fibre Channel command.

kB, kbyte Abbreviations for Kilobyte.

Key Usually a sequence of random or pseudorandom bits used to direct cryptographic operations and/or for producing other keys. The same plain text encrypted with different keys yields different ciphertexts, each of which requires a different key for decryption. In a symmetric cryptosystem, the encryption and decryption keys are the same. In an asymmetric cryptosystem, the encryption and decryption keys are different.

Key exchange A cryptographic protocol and procedure in which two communicating entities determine a shared key in a fashion such that a third party who reads all their communications cannot effectively determine the value of the key. A common approach to key exchange requires such a third party to compute a discrete logarithm over a large field in order to determine the key value and relies on the computational intractability of the discrete logarithm problem for suitably selected large fields for its security.

Key management The supervision and control of the process by which keys are generated, stored, protected, transferred, loaded, used, revoked, and destroyed.

Key pair A public key and its corresponding private key as used in public key cryptography (that is, asymmetric cryptosystem).

Keying material A key or authentication information in physical or magnetic form.

Kilobyte 1,000 (10^3) bytes of data. Common storage industry usage: 1,024 (2^{10}) bytes of data. Common usage in software contexts is typically clear from the context in which the term is used.

L_Port A port used to connect a node to a Fibre Channel arbitrated loop.

Label An identifier associated with a removable media or cartridge. Labels may be humanly readable, machine readable, or both.

LAN Local area network.

LANE Local area network emulation.

LAN-free backup A disk backup methodology in which a SAN appliance performs the actual backup I/O operations, thus freeing the LAN server to perform I/O operations on behalf of LAN clients. Differentiated from serverless backup by the requirement of an additional SAN appliance to perform the backup I/O operations.

Large read request, large write request, large I/O request An I/O request that specifies the transfer of a large amount of data. *Large* obviously depends on the context but typically refers to requests for 64 kB or more.

Latency I/O request execution time, the time between the making of an I/O request and completion of the request's execution; short for rotational latency, the time between the completion of a seek and the instant of arrival of the first block of data to be transferred at the disk's read/write head.

Latent fault A failure of a system component that has not been recognized because the failed aspect of the component has not been exercised since the occurrence of the failure. A field-developed medium defect on a disk surface is a latent fault until an attempt is made to read the data in a block that spans the defect.

LBA Logical block address.

LDAP Lightweight Directory Access Protocol.

LDM Logical disk manager.

LED Light-emitting diode.

Library A robotic media handler capable of storing multiple pieces of removable media and loading and unloading them from one or more drives in arbitrary order.

Light-emitting diode (LED) A multimode light source based on inexpensive optical diodes; available in a variety of wavelengths; 1300-nm wavelength is typical for data communications. The practical transfer rate limit for LEDs is 266 Mbps.

Lightweight Directory Access Protocol (LDAP) An IETF protocol for creating, accessing, and removing objects and data from a directory; it provides the ability to search, compare, add, delete, and modify directory objects, as well as modifying the names of these objects. It also supports bind, unbind, and abandon (cancel) operations for a session. LDAP got its name from its goal of being a simpler form of the Directory Access Protocol (DAP) from the X.500 set of standards.

Link A physical connection (electrical or optical) between two nodes of a network (Fibre Channel). Two unidirectional fibers transmitting in opposite directions and their associated transmitters and receivers (Fibre Channel). The full-duplex FC-0 level association between FC-1 entities in directly attached ports (Fibre Channel). The point-to-point physical connection from one element of a Fibre Channel fabric to the next.

LIP Loop initialization primitive.

LISM Loop initialization select master.

Load balancing The adjustment of system and/or application components and data so that application I/O or computational demands are spread as evenly as possible across a system's physical resources. I/O load balancing may be done manually (by a human) or automatically (by some means that does not require human intervention).

Load optimization The manipulation of an I/O load in such a way that performance is optimal by some objective metric. Load optimization may be achieved by load balancing across several components or by other means, such as request reordering or interleaved execution.

Glossary

Load sharing The division of an I/O load or task among several storage subsystem components without any attempt to equalize each component's share of the work. Each affected component bears a percentage of a shared load. When a storage subsystem is load sharing, it is possible for some of the sharing components to be operating at full capacity, to the point of actually limiting performance, whereas others are underutilized.

Local area network (LAN) A communications infrastructure designed to use dedicated wiring over a limited distance (typically a diameter of less than 5 km) to connect a large number of intercommunicating nodes. Ethernet and Token Ring are the two most popular LAN technologies.

Local area network emulation (LANE) A collection of protocols and services that combine to create an emulated LAN using ATM as the underlying network; LANE enables intelligent devices with ATM connections to communicate with remote LAN-connected devices as if they were directly connected to the LAN.

Local F_Port The F_Port to which a particular N_Port is directly attached by a link.

Logical block A block of data stored on a disk or tape and associated with an address for purposes of retrieval or overwriting. The term *logical block* is typically used to refer to the host's view of data addressing on a physical device. Within a storage device, there is often a further conversion between the logical blocks presented to hosts and the physical media locations at which the corresponding data are stored.

Logical block address (LBA) The address of a logical block. Logical block addresses are typically used in hosts' I/O commands. The SCSI Disk Command Protocol, for example, uses logical block addresses.

Logical disk A set of consecutively addressed FBA disk blocks that is part of a single virtual disk to physical disk mapping. Logical disks are used in some array implementations as constituents of logical volumes or partitions. Logical disks are normally not visible to the host environment, except during array configuration operations.

Logical disk manger (LDM)   A name for the volume management control software in the Windows NT operating system.

Logical unit   The entity within a SCSI target that executes I/O commands. SCSI I/O commands can be sent to a target and executed by a logical unit within that target. An SCSI physical disk typically has a single logical unit. Tape drives and array controllers may incorporate multiple logical units to which I/O commands can be addressed. Each logical unit exported by an array controller corresponds to a virtual disk.

Logical unit number (LUN)   The SCSI identifier of a logical unit within a target.

Logical volume   A virtual disk made up of logical disks; also called a virtual disk or volume set.

Login server   An intelligent entity within a Fibre Channel fabric that receives and executes fabric login requests.

Long-wavelength laser (LWL)   A laser with a wavelength 1,300 nm or longer (usually 1,300 or 1,550 nm) widely used in the telecommunications industry.

Loop initialization   The protocol by which a Fibre Channel arbitrated loop network initializes on power up or recovers after a failure or other unexpected condition; during LIP, the nodes present on the arbitrated loop identify themselves and acquire addresses on the loop for communication. No data can be transferred on an arbitrated loop until a LIP is complete.

Loop initialization primitive (LIP)   A Fiber Channel primitive used to initiate a procedure that results in unique addressing for all nodes, to indicate a loop failure, or to reset a specific node.

Loop initialization select master (LISM)   The process by which a temporary Fibre Channel arbitrated loop master is determined during loop initialization.

Loopback   FC-1 operational mode in which information passed to the FC-1 transmitter is shunted directly to the FC-1 receiver. When a Fibre Channel interface is in loopback mode, the loopback signal overrides any external signal detected by the receiver.

Glossary

Loop port state machine Logic that monitors and performs the tasks required for initialization and access to a Fibre Channel arbitrated loop.

LWL Long-wavelength laser.

LUN Logical unit number.

MAC Media access control.

Magnetic remanance A magnetic representation of residual information remaining on a magnetic medium after the medium has been degaussed.

MAN Metropolitan area network.

Managed Object Format (MOF) The syntax and formal description of the objects and associations in the CIM schemas; MOF also can be translated to XML using a document type definition published by the DMTF.

Management Information Base (MIB) The specification and formal description of a set of objects and variables that can be read and possibly written using the SNMP; various standard MIBs are defined by the IETF.

Mandatory (provision) Provision in a standard that must be supported in order for an implementation of the standard to be compliant.

Mapping Conversion between two data addressing spaces; for example, mapping refers to the conversion between physical disk block addresses and the block addresses of the virtual disks presented to operating environments by control software.

Mapping boundary A virtual disk block address of some significance to a disk array's mapping algorithms. The first and last blocks of a user data space strip or check data strip are mapping boundaries.

Maximum transfer unit (MTU) The largest amount of data that it is permissible to transmit as one unit according to a protocol specification; Ethernet MTU is 1,536 eight-bit bytes. The Fibre Channel MTU is 2,112 eight-bit bytes.

MB, Mbyte Abbreviations for megabyte.

Mb, Mbit Abbreviations for megabit.

MBps Megabytes per second. A measure of bandwidth or data transfer rate.

Mbps Megabits per second. A measure of bandwidth or data transfer rate.

MD5 A specific message-digest algorithm producing a 128-bit digest used as authentication data by an authentication service.

Mean time between failures (MTBF) The average time from start of use to first failure in a large population of identical systems, components, or devices.

Mean time to (loss of) data availability (MTDA) The average time from startup until a component failure causes a loss of timely user data access in a large population of storage devices. Loss of availability does not necessarily imply loss of data; for some classes of failures (for example, failure of nonredundant intelligent storage controllers), data remain intact and can again be accessed after the failed component is replaced.

Mean time to data loss (MTDL) The average time from startup until a component failure causes a permanent loss of user data in a large population of storage devices. Mean time to data loss is similar to MTBF for disks and tapes but is likely to differ in RAID arrays, where redundancy can protect against data loss due to component failures.

Mean time to repair (MTTR) The average time between a failure and completion of repair in a large population of identical systems, components, or devices. MTTR comprises all elements of repair time, from the occurrence of the failure to restoration of complete functionality of the failed component. This includes time to notice and respond to the failure, time to repair or replace the failed component, and time to make the replaced component fully operational. In mirrored and RAID arrays, for example, the mean time to repair a disk failure includes the time required for reconstructing user data and check data from the failed disk on the replacement disk.

Meaningful control field In a standard, a control field or bit that must be interpreted correctly by a receiver. Control fields are either meaningful or not meaningful, in which case they must be ignored.

Media access control (MAC) Algorithms that control access to physical media, especially in shared media networks.

Media ID A machine-readable identifier written on a piece of removable medium that remains constant throughout the medium's life.

Media manager A backup software component responsible for tracking the location, contents, and state of removable storage media.

Media robot Robotic media handler.

Media stacker A robotic media handler in which the medium must be moved sequentially by the robot, usually services a single drive. A stacker may be able to load media into a drive in arbitrary order but must cycle through media in sequence to do so.

Medium Material in a storage device on which data are recorded; a physical link on which data are transmitted between two points.

Megabaud One million bauds (elements of transmitted information) per second, including data, signaling, and overhead.

Megabit (Mb) 1,000,000 (10^6) bits. The SNIA uses the 10^6 convention commonly found in data transfer-related literature rather than 1,048,576 (2^{20}) bits, the convention common in the computer system RAM and software literature.

Megabyte (MB) 1,000,000 (10^6) bytes. The SNIA uses the 10^6 convention commonly found in the storage and data transfer-related literature rather than 1,048,576 (2^{20}) bits, the convention common in the computer system RAM and software literature.

Megatransfer The transfer of 1 million data units per second. Used to describe the characteristics of parallel I/O buses such as SCSI, for which the data transfer rate depends on the amount of data transferred in each data cycle.

Member, member disk A disk that is in use as a member of a disk array.

Message-digest algorithm An algorithm that produces a secure hash.

Metadata Data that describe data. In disk arrays, metadata consist of such items as array membership, member extent sizes and locations, descriptions of logical disks and partitions, and array state information. In file systems, metadata include file names, file properties and security information, and lists of block addresses at which each file's data are stored.

Metropolitan area network (MAN) A network that connects nodes distributed over a metropolitan (city-wide) area as opposed to a local area (campus) or wide area (national or global) network; from a storage perspective, MANs are of interest because there are MANs over which block storage protocols (for example, ESCON, Fibre Channel) can be carried natively, whereas most WANs that extend beyond a single metropolitan area do not currently support such protocols.

MIB Management information base.

MIME Multipurpose Internet Mail Extensions.

Mirroring A form of storage array in which two or more identical copies of data are maintained on separate media, known as RAID level 1, disk shadowing, real-time copy, and t1 copy.

Mirrors, mirrored disks The disks of a mirrored array.

Mirrored array Common term for a disk array that implements RAID level 1, or mirroring, to protecting data against loss due to disk or device I/O bus failure.

MLS Multilevel security.

Modeling language A language for describing the concepts of an information or data model. One of the most popular modeling languages in use today is Unified Modeling Language (UML). The essence of modeling languages is that they be capable of conveying the model concepts.

MOF Managed Object Format.

Monitor (program) A program that executes in an operating environment and keeps track of system resource utilization. Monitors typically record CPU utilization, I/O request rates, data transfer rates, RAM utilization, and similar statistics. A monitor program, which may be an integral part of an operating system, a

Glossary

separate software product, or a part of a related component, such as a database management system, is a necessary prerequisite to manual I/O load balancing.

Mount   In the Network File System (NFS), a protocol and set of procedures to specify a remote host and file system or directory to be accessed. Also specified is the location of the accessed directories in the local file hierarchy.

MTBF   Mean time between failures.

MTDA   Mean time until (loss of) data availability.

MTDL   Mean time to data loss.

MTTR   Mean time to repair.

MTU   Maximum transfer unit.

Multicast   The simultaneous transmission of a message to a subset of more than one of the ports connected to a communication facility. In a Fibre Channel context, multicast specifically refers to the sending of a message to multiple N_Ports connected to a fabric.

Multicast group   A set of ports associated with an address or identifier that serves as the destination for multicast packets or frames that are to be delivered to all ports in the set.

Multilevel disk array   A disk array with two levels of data mapping. The virtual disks created by one mapping level become the members of the second level. The most frequently encountered multilevel disk arrays use mirroring at the first level and stripe data across the resulting mirrored arrays at the second level.

Multilevel security (MLS)   Allows users and resources of different sensitivity levels to access a system concurrently while ensuring that only information for which the user or resource has authorization is made available. Requires a formal computer security policy model that assigns specific access characteristics to both subjects and objects.

Multimode (fiberoptic cable)   A fiberoptic cabling specification that allows up to 500m distances between devices.

Multithreaded   Having multiple concurrent or pseudoconcurrent execution sequences. Used to describe processes in computer systems. Multithreaded processes are one means by which I/O request-intensive applications can make maximum use of disk arrays to increase I/O performance.

Multipath I/O The facility for a host to direct I/O requests to a storage device on more than one access path. Multipath I/O requires that devices be uniquely identifiable by some means other than by bus address.

Multipurpose Internet Mail Extensions (MIME) A specification that defines the mechanisms for specifying and describing the format of Internet message bodies. An HTTP response containing a MIME content-type header allows the HTTP client to invoke the appropriate application for processing the received data.

N_Port Port that connects a node to a fabric or to another node; N_Ports connect to fabrics' F_Ports or to other nodes' N_Ports. N_Ports handle creation, detection, and flow of message units to and from the connected systems. N_Ports are end points in point-to-point links.

N_Port name A name identifier associated with an N_Port.

NAA Network Address Authority.

Name identifier A 64-bit identifier consisting of a 60-bit value concatenated with a 4-bit NAA identifier; name identifiers identify Fibre Channel entities such as N_Port, node, F_Port, or fabric.

Name server An intelligent entity in a network that translates between symbolic node names and network addresses. In a Fibre Channel network, a name server translates between worldwide names and fabric addresses.

Naming The mapping of address space to a set of objects; naming is typically used either for human convenience (for example, symbolic names attached to files or storage devices) or to establish a level of independence between two system components (for example, identification of files by iNode names or identification of computers by IP addresses).

Namespace The set of valid names recognized by a file system; one of four basic functions of file systems is maintenance of a

Glossary

namespace so that invalid and duplicate names do not occur. In XML, a document at a specific web address (URL) that lists the names of data elements and attributes that are used in other XML files; in CIM and WBEM, a collection of object definitions and instances that are logically consistent.

NAS Network attached storage.

National Committee on Information Technology (IT) Standards A committee of ANSI that serves as the governing body of X3T11 and other standards organizations.

National Institute of Standards and Technology (NIST) A nonregulatory federal agency within the U.S. Commerce Department's Technology Administration. NIST's mission is to develop and promote measurement, standards, and technology to enhance productivity, facilitate trade, and improve the quality of life. Specifically, the Computer Security Division within NIST's Information Technology (IT) Laboratory managed the Advanced Encryption Standard (AES) program.

NCITS National Committee on Information Technology (IT) Standards.

NDMP Network Data Management Protocol.

Network An interconnect that enables communication among a collection of attached nodes. A network consists of optical or electrical transmission media, infrastructure in the form of hubs and/or switches, and protocols that make message sequences meaningful. In comparison with I/O buses, networks typically are characterized by large numbers of nodes that act as peers, large internode separation, and flexible configurability.

Network adapter An adapter that connects an intelligent device to a network. Usually called a *network interface card* or *Ethernet NIC*.

Network address authority (NAA) (identifier) A 4-bit identifier defined in FC-PH to denote a network address authority (that is, an organization such as CCITT or IEEE that administers network addresses).

Network attached storage (NAS)   Reference to storage elements that connect to a network and provide file access services to computer systems; NAS storage element consists of an engine that implements the file services and one or more devices on which data are stored. NAS elements may be attached to any type of network. When attached to SANs, NAS elements may be considered to be members of the SAS class of storage elements; a class of systems that provide file services to host computers; a host system that uses network attached storage employs a file system device driver to access data using file access protocols such as NFS or CIFS. NAS systems interpret these commands and perform the internal file and device I/O operations necessary to execute them.

Network Data Management Protocol (NDMP)   Communications protocol that allows intelligent devices on which data are stored, robotic library devices, and backup applications to intercommunicate for the purpose of performing backups. An open standard protocol for network-based backup of NAS devices. NDMP allows a network backup application to control the retrieval of data from and backup of a server without third-party software. The control and data transfer components of backup and restore are separated. NDMP is intended to support tape drives but can be extended to address other devices and media in the future.

Network File System (NFS) Protocol   A distributed File System and its associated network protocol originally developed by Sun Microsystems Computer Corporation and commonly implemented in UNIX systems, although most other computer systems have implemented NFS clients and/or servers. The IETF is responsible for the NFS standard.

Network interface card (NIC)   An I/O adapter that connects a computer or other type of node to a network. NIC is usually a circuit module; however, the term is sometimes used to denote an ASIC or set of ASICs on a computer system board that perform the network I/O adapter function. The term *NIC* is used universally in Ethernet and Token Ring contexts. In Fibre Channel contexts, the terms *adapter* and *NIC* are used in preference to HBA.

NFS   Network File System.

NIC   Network interface card.

NIST   National Institute of Standards and Technology.

NL_Port   A port specific to Fibre Channel arbitrated loop; an NL_Port has the same functional, logical, and message-handling capability as an N_Port but connects to an arbitrated loop rather than to a fabric. Some implementations can function either as N_Ports or as NL_Ports depending on the network to which they are connected. An NL_Port must replicate frames and pass them on when in passive loop mode.

Node   An addressable entity connected to an I/O bus or network. Used primarily to refer to computers, storage devices, and storage subsystems. The component of a node that connects to the bus or network is a port.

Node name   A name identifier associated with a node.

Normal operation, normal mode   A state of a system in which the system is functioning within its prescribed operational bounds; for example, when a disk array subsystem is operating in normal mode, all disks are up, no extraordinary actions (for example, reconstruction) are being performed, and environmental conditions are within operational range. Sometimes called *optimal mode*.

Nonlinear mapping   Any form of tabular mapping in which there is not a fixed size correspondence between the two mapped address spaces. Nonlinear mapping is required in disk arrays that compress data because the space required to store a given range of virtual blocks depends on the degree to which the contents of those blocks can be compressed and therefore changes as block contents change.

Non-OFC (laser)   A laser transceiver whose lower-intensity output does not require special open fiber control (OFC) mechanisms.

Nonrepeating ordered set   An ordered set passed by FC-2 to FC-1 for transmission that has nonidempotent semantics; that is, it cannot be retransmitted.

Nonrepudiation   Assurance that a subject cannot later deny performance of some action; for communications, this may involve providing the sender of data with proof of delivery and the recipient with proof of the sender's identity so that neither can later deny having participated in the communication. Digital signatures are often used as a nonrepudiation mechanism for stored information in combination with timestamps.

Nontransparent fail-over A fail-over from one component of a redundant system to another that is visible to the external environment. For example, a controller fail-over in a redundant disk subsystem is nontransparent if the surviving controller exports the other's virtual disks at different host I/O bus addresses or on a different host I/O bus.

Nonuniform memory architecture (NUMA) A computer architecture that enables memory to be shared by multiple processors, with different processors having different access speeds to different parts of the memory.

Nonvolatile random access memory (NVRAM) Computer system random access memory that has been made impervious to data loss due to power failure through the use of UPS, batteries, or implementation technology such as flash memory.

Nonvolatility Property of data; nonvolatility refers to the property that data will be preserved even if certain environmental conditions are not met. Used to describe data stored on disks or tapes. If electrical power to these devices is cut, data stored on them is nonetheless preserved.

Not operational (receiver or transmitter) A receiver or transmitter that is not capable of receiving or transmitting an encoded bit stream based on rules defined by FC-PH for error control. For example, FC-1 is not operational during initialization.

NUMA Nonuniform memory architecture.

NVRAM Nonvolatile random access memory.

NVRAM cache A quantity of NVRAM used as a cache. NVRAM cache is particularly useful in RAID array subsystems, filers, database servers, and other intelligent devices that must keep track of the state of multistep I/O operations even if power fails during the execution of the steps.

Glossary

NVRAM card   A printer circuit module containing NVRAM.

Object   In the context of access control, an entity to which access is controlled and/or usage of which is restricted to authorized subjects. Information system resources are often examples of objects.

Object-oriented (OO) methodology   Methodology for decomposing an entity or problem by its key abstractions versus by its procedures or steps. The key abstractions become classes in an information or data model and embody well-defined behaviors called *methods* with a unique set of data attributes. Instances of a class are called *objects*.

OC-n   A data rate that is a multiple of the fundamental SONET rate of 51.84 Mbps. OC-3 (155 Mpbs), OC-12 (622 Mbps), OC-48 (2488 Mbps), and OC-192 (9953 Mbps) are currently in common use.

OFC   Open fiber control.

Offline backup   A backup created while the source data are not in use. Offline backups provide internally consistent Dictionary_C.html Term_checkpointimages of source data.

Online backup   A backup created while the source data are in use by applications. If source data are being modified by applications during an online backup, the backup is not guaranteed to be an internally consistent image of application data because of some point-in-time Dictionary_C.html Term_checkpointimages. Snapshots of online data initiated while applications are in a quiescent state are sometimes used as backup source data to overcome this limitation.

OO   Object-oriented.

Open   General: Any system or aspect of a system whose function is governed by a readily accessible standard rather than by a privately owned specification. Fibre Channel: A period of time that begins when a sequence or exchange is initiated and ends when the sequence or exchange is normally or abnormally terminated. General: Not electrically terminated, as in an unplugged cable.

Open fiber control (OFC)   A safety interlock system that limits the optical power level on an open optical fiber cable.

Open Group A cross-industry consortium for open systems standards and their certification. UNIX, management, and security standards are developed within the Open Group.

Open interconnect Standard interconnect.

Operating environment A collective term for the hardware architecture and operating system of a computer system.

Operation An FC-2 construct that encapsulates one or more possibly concurrent exchanges into a single abstraction for higher-level protocols.

Operation associator A value used in the association header to identify a specific operation within a node and correlate communicating processes related to that operation. The operation associator is a handle by which an operation within a given node is referred to by another communicating node. Operation associator is a generic reference to originator operation associator and responder operation associator.

Operational (state) The state of a receiver or transmitter that is capable of receiving or transmitting an encoded bit stream based on the rules defined by FC-PH for error control. Receivers capable of accepting signals from transmitters requiring laser safety procedures are not considered operational after power-on until a signal of duration longer than that associated with laser safety procedures is present at the fiber attached to the receiver.

Optical fall time The time interval required for the falling edge of an optical pulse to transition between specified percentages of the signal amplitude. For lasers, the transitions are measured between the 80 and 20 percent points. For LED media, the specification points are 90 and 10 percent.

Optional (characteristic) Characteristics of a standard that are specified by the standard but not required for compliance. If an optional characteristic of a standard is implemented, it must be implemented as defined in the standard.

Ordered set A transmission word (sequence of four 10-bit code bytes) with a special character in its first (leftmost) position and data characters in the remaining three positions. An ordered set is represented by the combination of special codes and data bytes

Glossary

that, when encoded, result in the generation of the transmission characters specified for the ordered set. Ordered sets are used for low-level Fibre Channel link functions such as frame demarcation, signaling between the ends of a link, initialization after power-on, and some basic recovery actions.

Originator The party initiating an exchange.

Originator exchange identifier (OX_ID) An identifier assigned by an originator to identify an exchange; an OX_ID is meaningful only to its originator.

Overwrite procedure The process of writing patterns of data on top of the data stored on a magnetic medium for the purpose of obliterating the data.

Out-of-band (transmission) Transmission of management information for Fibre Channel components outside the Fibre Channel network, typically over Ethernet.

Out-of-band virtualization Virtualization functions or services that are not in the data path. Examples are functions related to metadata, the management of data or storage, security management, backup of data, and so on.

OX_ID Originator exchange identifier.

Panic A colloquial term describing a software program's reaction to an incomprehensible state. In an operating system context, a panic is usually a system call or unexpected state that causes the system to abruptly stop executing so as to eliminate the possibility that the cause of the panic will cause further damage to the system, applications, or data.

Parallel access array A disk array model in which data transfer and data protection algorithms assume that all member disks operate in unison, with each participating in the execution of every application I/O request. A parallel access array is only capable of executing one I/O request at a time. True parallel access would require that an array's disks be rotationally synchronized. In actual practice, arrays approximate parallel access behavior. Ideal RAID level 2 and RAID level 3 arrays are parallel access arrays.

Parallel (transmission) Simultaneous transmission of multiple data bits over multiple physical lines.

Parity disk In a RAID level 3 or 4 array, the single disk on which the parity check data are stored.

Parity RAID A collective term used to refer to Berkeley RAID levels 3, 4, and 5.

Parity RAID array A RAID array whose data-protection mechanism is one of Berkeley RAID levels 3, 4, or 5.

Partition A subdivision of the capacity of a physical or virtual disk. Partitions are consecutively numbered ranges of blocks that are recognized by MS-DOS, Windows, and most UNIX operating systems; the type of extent used to configure arrays; a contiguously addressed range of logical blocks on a physical medium that is identifiable by an operating system via the partition's type and subtype fields. A partition's type and subtype fields are recorded on the physical medium and hence make the partition self-identifying.

Partitioning Presentation of the usable storage capacity of a disk or array to an operating environment in the form of several virtual disks whose aggregate capacity approximates that of the underlying physical or virtual disk. Partitioning is common in MS-DOS, Windows, and UNIX environments. Partitioning is useful with hosts that cannot support the full capacity of a large disk or array as one device. It also can be useful administratively, for example, to create hard subdivisions of a large virtual disk.

Passive copper A low-cost Fibre Channel connection that allows up to 13 m of copper cable lengths.

Passphrase A sequence of characters longer than the acceptable length of a password that is transformed by a password system into a virtual password of acceptable length.

Password A protected private alphanumeric string used to authenticate an identity or to authorize access to data.

Path The access path from a host computer to a storage device; the combination of device address and file system directory elements used to locate a file within a file system; any route through an interconnect that allows two devices to communicate; a sequence of computer instructions that performs a given function, such as I/O request execution.

Glossary

Path length General: The number of instructions (a rough measure of the amount of time) required by a computer to perform a specific activity, such as I/O request execution. Backup file system: The number of characters in a path name.

Path name The complete list of nested subdirectories through which a file is reached.

Payload Contents of the data field of a communications frame or packet; in Fibre Channel, the payload excludes optional headers and fill bytes, if they are present.

PB, Pbyte Petabyte (10^{15} bytes).

PBC Port bypass circuit.

PCI Peripheral component interconnect.

PDC Primary domain controller.

PCNFSD A daemon that permits personal computers to access file systems accessed through the NFS Protocol.

Penetration An unauthorized bypassing of the security mechanisms of a system.

Peripheral component interconnect (PCI) A bus for connecting interface modules to a computer system; variations of PCI support 32- and 64-bit parallel data transfers at 33- and 66-MHz cycle times; a 133-MHz PCIX has been proposed by Compaq, HP, and IBM.

Persistence Nonvolatility; usually used to distinguish between data and metadata held in DRAM, which is lost when electrical power is lost, and data held on nonvolatile storage (disk, tape, battery-backed DRAM, and so on) that survive or persists across power outages.

Physical configuration The installation, removal, or reinstallation of disks, cables, HBAs, and other components required for a system or subsystem to function. Physical configuration is typically understood to include address assignments, such as PCI slot number, SCSI target ID and logical unit number, and so on.

Physical block A physical area on a recording medium at which data are stored. Distinguished from the logical and virtual block views typically presented to the operating environment by storage devices.

Physical block address The address of a physical block. A number that can be algorithmically converted to a physical location on storage media.

Physical disk Storage system: a disk. Used to emphasize a contrast with virtual disks. Operating system: a host operating system's view of an online storage device.

Physical extent A number of consecutively addressed blocks on a physical disk. Physical extents are created by control software as building blocks from which redundancy groups and volume sets are created. Called p_extent by ANSI.

Physical extent block number The relative position of a block within a physical extent. Physical extent block numbers are used to develop higher-level constructs in RAID array striped data mapping, not for applications or data addressing.

PKI Public key infrastructure.

Plain text Unencrypted information.

PLDA Private loop direct attach.

PLOGI Port logic.

Point-in-time copy A fully usable copy of a defined collection of data that contains an image of the data as they appeared at a single point in time. The copy is considered to have logically occurred at that point in time, but implementations may perform part or all of the copy at other times (for example, via database log replay or rollback) as long as the result is a consistent copy of the data as they appeared at that point in time. Implementations may restrict point-in-time copies to be read-only or may permit subsequent writes to the copy. Three important classes of point-in-time copies are split mirror, changed block, and concurrent. Pointer remapping and copy on write are implementation techniques often used for the latter two classes.

Pointer copy A point-in-time copy made using the pointer remapping technique.

Pointer remapping A technique for maintaining a point-in-time copy in which pointers to all the source data and copy data are maintained. When data are overwritten, a new location is chosen for the updated data, and the pointer for that data is remapped to point to it. If the copy is read-only, pointers to its data are never modified.

Policy processor In an intelligent device, the processor that schedules the overall activities. Additional processors, state machines, usually augment policy processors or sequencers that perform the lower-level functions required for implementing overall policy.

Port I/O adapter used to connect an intelligent device (node) to an I/O bus or network. Storage subsystems: The head end of a device I/O bus containing the arbitration logic.

Port_ID A unique 24-bit address assigned to an N_Port. There may be at most 2^{24} or 16.7 million N_Ports in a single Fibre Channel fabric. There may be at most 127 NL_Ports in a single loop. For point-to-point (N_Port-to-N_Port) connections, there are only 2. In some implementations, a device's Port_ID is derived from its worldwide name. In other cases, Port_IDs are permanently assigned in association with specific physical ports. Port_IDs may or may not survive a loop initialization process or, in the case of a switched fabric, a reconfiguration of the Fibre Channel switches.

Port bypass circuit A circuit that automatically opens and closes a Fibre Channel arbitrated loop so that nodes can be added to or removed from the loop with minimal disruption of operations. Port bypass circuits typically are found in Fibre Channel hubs and disk enclosures.

Port login The port-to-port login process by which Fibre Channel initiators establish sessions with targets.

Port name A unique 64-bit identifier assigned to a Fibre Channel port.

POST Power-on self-test.

Power conditioning The regulation of power supplied to a system so that acceptable ranges of voltage and frequency are maintained. Power conditioning is sometimes done by a storage subsystem but also may be an environmental requirement.

Power-on self-test (POST)   A set of internally stored diagnostic programs run by intelligent devices when powered on; these diagnostic programs verify the basic integrity of hardware before software is permitted to run on it.

Present (verb)   To cause to appear or to make available; for example, RAID control software and volume managers present virtual disks to host environments.

Primary domain controller (PDC)   A domain controller that has been assigned as or has negotiated to become the primary authentication server for the domain of which it is a part.

Primitive sequence   An ordered set transmitted repeatedly and continuously until a specified response is received.

Primitive signal   An ordered set with a special meaning such as an idle or receiver ready (R_RDY).

Private key   A key that is used in a symmetric cryptosystem in both encryption and decryption processes or in an asymmetric cryptosystem for one but not both of those processes. A private key must remain confidential to the using party if communication security is to be maintained.

Private key cryptography   An encryption methodology in which the encryptor and decryptor use the same key, which must be kept secret.

Private loop   A Fibre Channel arbitrated loop with no fabric attachment.

Private loop device   A Fibre Channel arbitrated loop device that does not support fabric login.

Process policy   An error-handling policy that allows an N_Port to continue processing data frames following detection of one or more missing frames in a sequence.

Process associator   A value in the association header that identifies a process or a group of processes within a node. Communicating processes in different nodes use process associators to address each other. Originating processes have originator process associators; responding processes have responder process associators.

Profile A proper subset of a standard that supports interoperability across a set of products or in a specific application. Profiles exist for FCP (FCSI and PLDA), IP, and other areas. A profile is a vertical slice through a standard containing physical, logical, and behavioral elements required for interoperability.

Proprietary interconnect, proprietary I/O bus An I/O bus (either a host I/O bus or a device I/O bus) whose transmission characteristics and protocols are the intellectual property of a single vendor and that requires the permission of that vendor to be implemented in the products of other vendors.

Protected space, protected space extent The storage space available for application data in a physical extent that belongs to a redundancy group.

Protocol A set of rules for using an interconnect or network so that information conveyed on the interconnect can be interpreted correctly by all parties to the communication. Protocols include such aspects of communication as data representation, data item ordering, message formats, message and response sequencing rules, block data transmission conventions, timing requirements, and so forth.

Public key A key that is used in an asymmetric cryptosystem for either the encryption or decryption process in which the private key is not used and which can be shared among a group of users without affecting the security of the cryptosystem.

Public key cryptography An encryption system using a linked pair of keys. What one key of the pair encrypts the other decrypts. Either key can be used for encryption and decryption.

Public key infrastructure (PKI) A framework established to issue, maintain, and revoke public key certificates accommodating a variety of security technologies.

Public loop A Fibre Channel arbitrated loop with an attachment to a fabric.

Public loop device A Fibre Channel arbitrated loop device that supports fabric login and services.

Pull technology The transmission of information in response to a request for that information. An example of a pull technology is polling.

Push technology The transmission of information from a source or initiator without the source being requested to send that information. An example of a push technology is an SNMP trap.

PVC Permanent virtual circuit.

QoS Quality of service.

Quiesce (verb) To bring a device or an application to a state where (1) it is able to operate, (2) all its data are consistent and stored on nonvolatile storage, and (3) processing has been suspended and there are no tasks in progress (that is, all application tasks have either been completed or not started).

Quiescent state An application or device state where (1) the application or device is able to operate, (2) all its data are consistent and stored on nonvolatile storage, and (3) processing has been suspended and there are no tasks in progress (that is, all tasks have either been completed or not started).

Quality of service (QoS) A technique for managing computer system resources such as bandwidth by specifying user-visible parameters such as message delivery time. Policy rules are used to describe the operation of network elements to make these guarantees. Relevant standards for QoS in the IETF are the Resource Reservation Protocol (RSVP) and Common Open Policy Service (COPS) Protocols. RSVP allows for the reservation of bandwidth in advance, whereas COPS allows routers and switches to obtain policy rules from a server.

RAID Redundant array of independent disks, a family of techniques for managing multiple disks to deliver desirable cost, data availability, and performance characteristics to host environments; a phrase adopted from the 1988 SIGMOD paper, "A Case for Redundant Arrays of Inexpensive Disks."

RAID 0, RAID level 0 Data striping.

RAID 1, RAID level 1 Mirroring.

RAID 2, RAID level 2 A form of RAID in which a Hamming code computed on stripes of data on some of an array's disks is stored on the remaining disks and serves as check data.

RAID 3, RAID level 3 A form of parity RAID in which all disks are assumed to be rotationally synchronized and where the data stripe size is no larger than the exported block size.

RAID 4, RAID level 4 A form of parity RAID in which the disks operate independently, the data strip size is no smaller than the exported block size, and all parity check data are stored on one disk.

RAID 5, RAID level 5 A form of parity RAID in which the disks operate independently, the data strip size is no smaller than the exported block size, and parity check data are distributed across the array's disks.

RAID 6, RAID level 6 Any form of RAID that can continue to execute read and write requests to all an array's virtual disks in the presence of two concurrent disk failures. Both dual check data computations (parity and Reed Solomon) and orthogonal dual parity check data have been proposed for RAID level 6.

RAID Advisory Board An organization of suppliers and users of storage subsystems and related products whose goal is to foster the understanding of storage subsystem technology among users and to promote all aspects of storage technology in the market.

RAMdisk A quantity of host system RAM managed by software and presented to applications as a high-performance disk. RAMdisks generally emulate disk I/O functional characteristics, but unless augmented by special hardware to make their contents nonvolatile, they cannot tolerate loss of power without losing data.

Random I/O, Random I/O load, Random reads, Random writes Any I/O load whose consecutively issued read and/or write requests do not specify adjacently located data. The term *random I/O* is used commonly to denote any I/O load that is not sequential, whether or not the distribution of data locations is indeed random. Random I/O is characteristic of I/O request-intensive applications.

Random relative offset A transmission control algorithm in which the frames containing the subblocks that comprise a block of information may be transmitted in any order. This complicates reassembly and detection of lost frames by comparison with continuously increasing relative offset.

Rank A set of physical disk positions in an enclosure, usually denoting the disks that are or can be members of a single array; the sets of corresponding target identifiers on all a controller's device I/O buses. Like the preceding definition, the disks identified as a rank by this definition usually are or can be members of a single array, a stripe in a redundancy group. Because of the diversity of meanings attached to this term by disk subsystem developers, SNIA publications make minimal use of it.

RAS Reliability, availability, and serviceability; remote access server (Windows NT dialup networking server).

Raw partition A disk partition not managed by a volume manager. The term *raw partition* is frequently encountered when discussing database systems because some database system vendors recommend volumes or files for underlying database storage, whereas others recommend direct storage on raw partitions.

Raw partition backup A bit-by-bit copy of a partition image. A raw partition backup incorporates no information about the objects contained on the partition and hence cannot be used for individual object restoration.

Read/write head The magnetic or optical recording device in a disk. Read/write heads are used both to write data by altering the recording medium's state and to read data by sensing the alterations. Disks typically have read/write heads, unlike tapes, where reading and writing are often done using separate heads.

Real-time copy Mirroring.

Rebuild, rebuilding Regeneration and writing onto one or more replacement disks of all the user data and check data from a failed disk in a mirrored or RAID array. In most arrays, a rebuild can occur while applications are accessing data on the array's virtual disks.

Receiver An interconnect or network device that includes a detector and signal-processing electronics. The portion of a link control facility dedicated to receiving an encoded bit stream, converting the stream into transmission characters, and decoding the characters using the rules specified by FC-PH; a circuit that converts an optical or electrical media signal to an (possibly retimed) electrical serial logic signal.

Receptacle The stationary (female) half of the interface connector on a transmitter or receiver.

Reconstruction Rebuilding.

Recorded volume serial number (RVSN) Media ID.

Recovery The recreation of a past operational state of an entire application or computing environment. Recovery is required after an application or computing environment has been destroyed or otherwise rendered unusable. It may include restoration of application data if that data have been destroyed as well.

Red In the context of security analysis, a designation applied to information systems and associated areas, circuits, components, and equipment where sensitive information is being processed.

Red/black concept The separation of electrical and electronic circuits, components, equipment, and systems that handle sensitive information (red) in electrical form from those that handle on information that is not sensitive (black) in the same form.

Reduced mode Degraded mode.

Reduction The removal of a member disk from a RAID array, placing the array in degraded mode. Reduction most often occurs because of member disk failure; however, some RAID implementations allow reduction for system management purposes.

Redundancy The inclusion of extra components of a given type in a system (beyond those required by the system to carry out its function) for the purpose of enabling continued operation in the event of a component failure.

Redundancy group A collection of extents organized for the purpose of providing data protection. Within a redundancy group, a single type of data protection is employed. All the usable storage capacity in a redundancy group is protected by check data stored within the group, and no usable storage external to a redundancy group is protected by check data within it. A class defined in the CIM schema (CIM_RedundancyGroup) consisting of a collection of objects for which redundancy is provided. Three subclasses of CIM_RedundancyGroup are defined: (1) CIM_SpareGroup for sparing and fail-over, (2) CIM_ExtraCapacityGroup for load sharing or load balancing, and (3) CIM_StorageRedundancyGroup to describe the redundancy algorithm in use.

Redundancy group stripe A set of sequences of correspondingly numbered physical extent blocks in each of the physical extents comprising a redundancy group. The check data blocks in a redundancy group stripe protect the protected space in that stripe.

Redundancy group stripe depth The number of consecutively numbered physical extent blocks in one physical extent of a redundancy group stripe. In the conventional striped data mapping model, redundancy group stripe depth is the same for all stripes in a redundancy group.

Redundant (components) Components of a system that have the capability to substitute for each other when necessary, such as, for example, when one of the components fails, so that the system can continue to perform its function. In storage subsystems, power distribution units, power supplies, cooling devices, and controllers are often configured to be redundant. The disks comprising a mirror set are redundant. A parity RAID array's member disks are redundant because surviving disks can collectively replace the function of a failed disk.

Redundant (configuration, system) A system or configuration of a system in which failure tolerance is achieved by the presence of redundant instances of all components that are critical to the system's operation.

Glossary

Redundant array of independent disks (RAID)   A disk array in which part of the physical storage capacity is used to store redundant information about user data stored on the remainder of the storage capacity. The redundant information enables regeneration of user data in the event that one of the array's member disks or the access path to it fails. Although it does not conform to this definition, disk striping is often referred to as RAID (RAID level 0).

Regeneration   Recreation of user data from a failed disk in a RAID array using check data and user data from surviving members; regeneration also may be used to recover data from an unrecoverable media error. Data in a parity RAID array is regenerated by computing the exclusive OR of the contents of corresponding blocks from the array's remaining disks. Choosing the more convenient of two parity algorithms and executing it regenerates data in a RAID level 6 array.

Registered state change notification   A Fibre Channel switch function that allows notification to registered nodes if a change occurs to other specified nodes.

Rekeying   The process of changing the key used for an ongoing communication session.

Relative offset   A displacement, expressed in bytes, used to divide a quantity of data into blocks and subblocks for transmission in separate frames. Relative offsets are used to reassemble data at the receiver and verify that all data have arrived.

Relative offset space   A numerical range defined by a sending upper-level protocol for an information category. The range starts at zero, representing the upper-level defined origin, and extends to a highest value. Relative offset values are required to lie within the appropriate relative offset space.

Removable media storage device   A storage device designed so that its storage media can be readily removed and inserted. Tapes, CD-ROMs, and optical disks are removable media devices.

Repeater   A circuit that uses clock recovered from an incoming signal to generate an outbound signal.

Repeating ordered set An ordered set issued by FC-2 to FC-1 for repeated transmission until a subsequent transmission request is issued by FC-2.

Replacement disk A disk available for use as or used to replace a failed member disk in a RAID array.

Replacement unit (RU) A component or collection of components in a system that are always replaced (swapped) as a unit when any part of the collection fails. Replacement units may be field-replaceable, or they may require that the system of which they are part be returned to a factory or repair depot for replacement. Field-replaceable units may be customer-replaceable, or their replacement may require trained service personnel. Typical replacement units in a disk subsystem include disks, controller logic boards, power supplies, cooling devices, and cables. Replacement units may be cold, warm, or hot swappable.

Replay attack An attack in which a valid data transmission is maliciously or fraudulently repeated, either by the originator or by an adversary who intercepts the data and retransmits them.

Replica A general term for a copy of a collection of data.

Replicate Noun: A general term for a copy of a collection of data. Verb: Action of making a replicate as defined above.

Request-intensive application A characterization of applications. Also known as *throughput-intensive*. A request-intensive application is an I/O-intensive application whose I/O load consists primarily of large numbers of I/O requests for relatively small amounts of data. Request-intensive applications are typically characterized by random I/O loads.

Request for comment (RFC) Internet-related specifications, including standards, experimental definitions, informational documents, and best-practice definitions produced by the IETF.

Request-intensive (application) A characterization of applications. A request-intensive application is an I/O-intensive application characterized by a high rate of I/O requests. Request-intensive applications' I/O requests are usually randomly addressed and often specify a small amount of data for transfer.

Glossary

Reserved (field) In a standard, a field in a data structure set aside for future definition. Some standards prescribe implementation behavior with respect to reserved fields (for example, originators of data structures containing reserved fields must zero fill them, consumers of data structures containing reserved fields must ignore them, and so on); others do not. A field filled with binary zeros by a source N_Port and ignored by a destination N_Port. Each bit in a reserved field is denoted by *r* in the Fibre Channel standards. Future enhancements to FC-PH may define usages for reserved fields. Implementations should not check or interpret reserved fields. Violation of this guideline may result in loss of compatibility with future implementations that comply with future enhancements to FC-PH.

Responder Target. Used only in Fibre Channel contexts.

Responder exchange identifier (RX_ID) An identifier assigned by a responder to identify an exchange; an RX_ID is meaningful only to the responder that originates it.

Restoration The copying of a backup to online storage for application use; restoration normally occurs after part or all of an application's data have been destroyed or become inaccessible.

Retention period Backup: The length of time that a backup image should be kept. File system: In some file systems, such as that shipped with IBM Corporation's OS/390 operating system, a property of a file that can be used to implement backup and data migration policies.

Retimer A circuit that uses a clock independent of the incoming signal to generate an outbound signal.

Return loss The ratio of the strength of a returned signal to that of the incident signal that caused it. In electric circuits, return loss is caused by impedance discontinuities. Optical return loss is caused by index of refraction differences.

RFC Request for comment.

Robot, robotic media handler A mechanical handler capable of storing multiple pieces of removable media and loading and unloading them from one or more drives in arbitrary order in response to electronic commands.

Rotational latency The interval between the end of a disk seek and the time at which the starting block address specified in the I/O request passes the disk head. Detailed disk drive simulation or measurement can only obtain exact rotational latencies for specific sequences of I/O operations. The simplifying assumption that, on average, requests wait for half a disk revolution time of rotational latency works well in practice. Half a disk revolution time is therefore defined to be the average rotational latency.

Row The set of blocks with corresponding physical extent block addresses in each of an array's member physical extents. The concept of rows is useful for locking the minimal amount of data during a RAID array update so as to maximize the potential for parallel execution.

RSA Both a public key algorithm and a corporation in the business of algorithm design, derived from the names of the founders (Rivest, Shamir, and Adelman).

RSCN Registered state change notification.

RU Replaceable unit.

Run length The number of consecutive identical bits in a transmitted signal. For example, the pattern 0011111010 has run lengths of 2, 5, 1, 1, and 1.

Running disparity The cumulative disparity (positive or negative) of all previously issued transmission characters.

RVSN Recorded volume serial number.

S_ID Source identifier.

S_Port A logical port inside a switch addressable by external N_Ports for service functions. An S_Port may be an implicit switch port or a separate entity such as a name server connected to and controlled by the switch. S_Ports have well-known port names to facilitate early discovery by N_Ports.

SAN Storage area network; this is the normal usage in SNIA documents; server area network that connects one or more servers; system area network for an interconnected set of system elements.

SAN attached storage (SAS) A term used to refer to storage elements that connect directly to a storage area network and provide file, database, block, or other types of data access services to computer systems; SAS elements that provide file access services are commonly called network attached storage (NAS) devices.

SAS SAN attached storage.

Saturated disk A disk whose instantaneous I/O load is as great as or greater than its capability to satisfy the requests comprising the load. Mathematically, a saturated disk's I/O queue eventually becomes indefinitely long. In practice, however, user reaction or other system factors generally reduce the rate of new request arrivals for a saturated disk.

Scale (verb) In computer systems, to grow or support growth in such a way that all capabilities of the system remain in constant ratio to each other. For example, a storage subsystem whose data transfer capacity increases by the addition of buses as its storage capacity increases by the addition of disks is said to scale.

Schema A collection of information models or data models.

Script A parameterized list of primitive I/O bus operations intended to be executed in sequence. Often used with respect to ports, most of which are able to execute scripts of I/O commands autonomously without policy processor assistance. A sequence of instructions intended to be parsed and carried out by another program. Perl, VBScript, JavaScript, and Tcl are all scripting languages.

SCSI Small Computer System Interface.

SCSI adapter An adapter that connects an intelligent device to an SCSI bus.

SCSI address The full address used by a computer to communicate with an SCSI device, including an adapter number (required with computers configured with multiple SCSI adapters) and the target ID of the device. SCSI addresses do not include logical unit number because those are not used for communication.

SCSI bus Any parallel (multisignal) I/O bus that implements some version of the ANSI SCSI standard. A wide SCSI bus may connect up to 16 initiators and targets. A narrow SCSI bus may connect up to 8 initiators and targets.

SCSI Enclosure Services (SES) An ANSI X3T10 standard for management of environmental factors such as temperature, power, voltage, and so on.

SCSI parallel interface (SPI) The family of SCSI standards that define the characteristics of the parallel version of the SCSI interface. Several versions of SPI, known as SPI, SPI2, SPI3, and so on, have been developed. Each version provides for greater performance and functionality than preceding ones.

SCSI Trade Association (STA) A trade association incorporated in 1996 to promote all forms of SCSI technology in the market.

SDH Synchronous Digital Hierarchy.

Sector The unit where data is physically stored and protected against errors on a fixed block architecture disk; a sector typically consists of a synchronization pattern, a header field containing the block's address, data, a checksum or error-correcting code, and a trailer. Adjacent sectors are often separated by information used to assist in track centering. Most often each sector holds a block of data.

Secure hash An algorithm that generates a digest from its input (for example, a message). The digest has the properties that different inputs are extraordinarily unlikely to have the same fingerprint, small changes in its input lead to large changes in its output, and it is computationally intractable to generate an input that has the same fingerprint as a given input.

Secure Sockets Layer (SSL) A suite of cryptographic algorithms, protocols, and procedures used to provide security for communications used to access the World Wide Web (WWW). The characters *https:* at the front of a Uniform Resource Locator (URL) cause SSL to be used to enhance communications security. More recent versions of SSL are known as Transport Level Security (TLS) and are standardized by the Internet Engineering Task Force (IETF).

SEQ_ID A number transmitted with each data frame in a sequence that identifies the frame as part of the sequence.

Glossary

Sequence A set of Fibre Channel data frames with a common sequence ID (SEQ_ID) corresponding to one message element, block, or information unit. Sequences are transmitted from initiator to recipient, with an acknowledgment, if applicable, transmitted from recipient to initiator.

Sequence initiative A Fibre Channel protocol feature that designates which end of an exchange has authority to send the next sequence.

Sequence initiator An N_Port that initiates a sequence and transmits data frames to a destination N_Port.

Sequence recipient (SR) An N_Port or NL_Port that receives data frames from a sequence initiator and, if applicable, transmits responses (link control frames) to the sequence initiator.

Sequence status block A data structure that tracks the state of a sequence. Both sequence initiators and sequence recipients have sequence status blocks for each active sequence.

Sequential I/O, Sequential I/O load, Sequential reads, Sequential writes An I/O load consisting of consecutively issued read or write requests to adjacently located data. Sequential I/O is characteristic of data transfer-intensive applications.

SERDES Serializer deserializer.

Serial (transmission) The transmission of data bits one at a time over a single link.

Serial adapter An adapter that connects an intelligent device to an RS232 or RS425 serial communications link, storage subsystems, filers, and other intelligent devices to connect to serial consoles for management purposes sometimes use serial adapters.

Serial console A real or emulated communication terminal used by humans to manage an intelligent device. Serial consoles connect to the devices' serial adapters.

Serial SCSI Any implementation of SCSI that uses serial data transmission (as opposed to multiconductor parallel buses). Optical and electrical Fibre Channel, SSA, and IEEE 1394 are examples of serial SCSI implementations.

Serializer deserializer (SERDES) A mechanism for converting data from parallel to serial form and from serial to parallel form.

Server An intelligent device, usually a computer, that provides services to other intelligent devices, usually other computers or appliances; an asymmetric relationship with a second party (a client) in which the client initiates requests and the server responds to those requests.

Server-based virtualization Virtualization implemented in a host computer rather than in a storage subsystem or storage appliance. Virtualization can be implemented either in host computers, in storage subsystems or storage appliances, or in a specific virtualization appliances in the storage interconnect fabric.

Serverless backup A disk backup methodology in which either the disk being backed up or the tape device receiving the backup manages and performs actual backup I/O operations. Server-free backup frees the LAN server to perform I/O operations on behalf of LAN clients and reduces the number of trips the backup data take through processor memory. Differentiated from LAN-free backup in that no additional SAN appliance is required to offload backup I/O operations from the LAN server.

Server Message Block (SMB) Protocol A network file system access protocol designed and implemented by Microsoft Corporation and used by Windows clients to communicate file access requests to Windows servers; current versions of the SMB Protocol are usually referred to as the Common Internet File System (CIFS).

Service level agreement (SLA) An agreement between a service provider, such as an IT department, an Internet services provider, or an intelligent device acting as a server, and a service consumer. A service level agreement defines parameters for measuring the service and states quantitative values for those parameters.

SES SCSI Enclosure Services; Solution Exchange Standard.

Share A resource such as data or a printer device made available for use by users on other computer systems. For example, a printer or a collection of files stored in a single directory tree on a file

Glossary

server may be made available as a share. CIFS clients, who include most networked personal computers, typically map a share to a drive letter.

Shielded enclosure A room or container designed to attenuate electromagnetic radiation.

Shelf A modular enclosure for storage devices (disks and tapes); storage shelves usually contain power supplies and cooling devices and have prewired backplanes that carry power and I/O bus signals to the devices mounted in them.

SIA Semiconductor Industries Association; SCSI Industry Association.

Simple name server (SNS) A service provided by a Fibre Channel switch that simplifies discovery of devices attached to the fabric.

SID Security identifier.

Simple Network Management Protocol (SNMP) An IETF protocol for monitoring and managing systems and devices in a network; the data being monitored and managed are defined by an MIB. The functions supported by the protocol are the request and retrieval of data, the setting or writing of data, and traps that signal the occurrence of events.

Single (component) configuration A configuration in which the referenced component is not redundant.

Single-ended (signaling) An electrical signaling technique in which all control and data signals are represented by a voltage difference from a common ground.

Single mode (fiberoptic cable) A fiberoptic cabling specification that provides for up to 10 km distance between devices.

Single point of failure (SPOF) One component or path in a system, the failure of which would make the system inoperable.

SIS Service Incident Standard.

SLA Service level agreement.

Small Computer System Interface (SCSI) A collection of ANSI standards and proposed standards that define I/O buses primarily intended for connecting storage subsystems or devices to hosts through HBAs. Originally intended primarily for use with small (desktop and deskside workstation) computers, SCSI has been extended to serve most computing needs and is arguably the most widely implemented I/O bus in existence today.

Small read request, small write request, small I/O request An I/O, read, or write request that specifies the transfer of a relatively small amount of data. *Small* usually depends on the context but most often refers to 8 kB or fewer.

SMB Server Message Block.

SMI Structure of management information.

SMPTE Society of Motion Picture and Television Engineers.

Snapshot A fully usable copy of a defined collection of data that contains an image of the data as they appeared at the point in time at which the copy was initiated. A snapshot may be either a duplicate or a replicate of the data it represents. The CIM_Snapshot class. An optional construct that can be used to represent a storage extent that contains either a full copy of another storage extent or the changes to that extent (in the case of a delta-before or delta-after copy). A CIM snapshot is not equivalent to a volume- or file-based snapshot or a point-in-time copy. It represents storage used to hold a copied image of an extent or to hold changes to an extent.

SNIA Storage Networking Industry Association.

Sniffer A software tool for auditing and identifying network traffic packets.

SNMP Simple Network Management Protocol.

SNS Simple name server.

Society of Motion Picture and Television Engineers (SMPTE) An industry association whose goal is to standardize television and motion picture industry information interchange protocols.

Soft zone A zone consisting of zone members that are permitted to communicate with each other via the fabric. Soft zones are typically implemented through a combination of name server and Fibre Channel protocol: When a port contacts the name server, the

Glossary

name server returns information only about Fibre Channel ports in the same zone(s) as the requesting port. This prevents ports outside the zone(s) from being discovered, and hence the Fibre Channel protocol will not attempt to communicate with such ports. In contrast to hard zones, soft zones are not enforced by hardware; for example, a frame that is erroneously addressed to a port that should not receive it will nonetheless be delivered. Well-known addresses {link} are implicitly included in every zone.

SOF   Start of frame.

Solicited control   An information category indicated in a Fibre Channel frame header.

Solicited data   An information category indicated in a Fibre Channel frame header.

Solid-state disk   A disk whose storage capability is provided by solid-state random access memory rather than by magnetic or optical media. A solid-state disk generally offers very high access performance compared with that of rotating magnetic disks because it eliminates mechanical seek and rotation time. It also may offer very high data transfer capacity. Cost per byte of storage, however, is typically quite high, and volumetric density is lower. A solid-state disk includes some mechanism such as battery backup or magnetic backing store that allows its operating environment to treat it as nonvolatile storage.

Solution Exchange Standard (SES)   A DMTF standard that defines the exchange of support or help desk information.

Service Incident Standard   A DMTF standard that defines how a support or help desk incident is processed.

SONET   Synchronous Optical Network.

Source identifier (S_ID)   A number in a Fibre Channel frame that identifies the source of the frame.

Source N_Port   The N_Port from which a frame is transmitted.

Spare (disk, extent)   An object reserved for the purpose of substitution for a like object in case of that object's failure.

Special character Any transmission character that is valid in the transmission code but does not correspond to a valid data byte. Special characters are used to denote special functions.

Special code A code that, when encoded using the rules specified by the transmission code, results in a special character. Special codes are typically associated with control signals related to protocol management (for example, K28.5).

SPI SCSI parallel interface.

Spiral data transfer rate Full-volume transfer rate.

Split I/O request An I/O request to a virtual disk that requires two or more I/O operations to satisfy because the virtual data addresses in the request map to more than one extent on one or more disks. An application I/O request that is divided into two or more subrequests by a file system or other operating system component because the amount of data requested is too large for the operating environment to handle as a unit.

Split mirror, split mirror copy, split mirror point-in-time copy Any of a class of point-in-time copy implementations or the resulting copies where the storage for the copy is synchronized to the source of the copy and then split. A split mirror copy occupies as much storage as the source of the copy.

SPOF Single point of failure.

Spoofing Unauthorized use of legitimate identification and authentication data to mimic a subject different from the attacker. Impersonating, masquerading, piggybacking, and mimicking are forms of spoofing.

SR Sequence recipient.

SRM Storage resource management.

STA SCSI Trade Association.

Stand-alone drive A removable media drive that is not associated with a media stacker or robot.

Standard interconnect An I/O or network interconnect whose specifications are readily available to the public and which can therefore be implemented easily in a vendor's products without license or royalty payments. Also called *open interconnect*.

Glossary

Star A physical network configuration in which every node is connected directly to and only to a central point; all communications pass through the central point, which may be a hub or a switch.

Start of frame (SOF) A group of ordered sets that delineates the beginning of a frame.

Storage area network (SAN) A network whose primary purpose is the transfer of data between computer systems and storage elements and among storage elements. A SAN consists of a communication infrastructure, which provides physical connections, and a management layer, which organizes the connections, storage elements, and computer systems so that data transfer is secure and robust. The term *SAN* is usually (but not necessarily) identified with block I/O services rather than file access services. A storage system consisting of storage elements, storage devices, computer systems, and/or appliances, plus all control software, communicating over a network. *Note:* The SNIA definition specifically does not identify the term *SAN* with Fibre Channel technology. When the term *SAN* is used in connection with Fibre Channel technology, use of a qualified phrase such as "Fibre Channel SAN" is encouraged. According to this definition, an Ethernet-based network whose primary purpose is to provide access to storage elements would be considered a SAN. SANs are sometimes also used for system interconnection in clusters.

Storage array A collection of disks or tapes from one or more commonly accessible storage subsystems combined with a body of control software.

Storage controller An intelligent controller to which storage devices are attached.

Storage device A collective term for disks, tapes, disk arrays, tape arrays, and any other mechanisms capable of nonvolatile data storage. This definition is specifically intended to exclude aggregating storage elements such as RAID array subsystems, robotic tape libraries, filers, and file servers.

Storage device virtualization Virtualization of storage devices such as disks, tape drives, RAID shelves, and so on.

Storage domain A collection of storage resources and supporting software and interfaces that are managed as a unit.

Storage element Any device designed and built primarily for the purpose of persistent data storage and delivery. This definition is specifically intended to encompass disk drives, tape drives, RAID array subsystems, robotic tape libraries, filers, file servers, and any other types of storage devices.

Storage extent A CIM object called CIM_StorageExtent. A storage extent instance may represent either removable or nonremoveable medium.

Storage networking The practice of creating, installing, administering, or using networks whose primary purpose is the transfer of data between computer systems and storage elements and among storage elements.

Storage Networking Industry Association (SNIA) An association of producers and consumers of storage networking products whose goal is to further storage networking technology and applications.

Storage resource management (SRM) Management of physical and logical storage resources, including storage elements, storage devices, appliances, virtual devices, disk volume, and file resources.

Storage subsystem An integrated collection of (1) storage controllers and/or HBAs, (2) storage devices such as disks, CDROMs, tapes, media loaders, and robots, and (3) any required control software that provides storage services to one or more computers.

Storage subsystem virtualization The implementation of virtualization in a storage subsystem.

Storage virtualization The act of abstracting, hiding, or isolating the internal function of a storage (sub)system or service from applications, compute servers, or general network resources for the purpose of enabling application and network-independent management of storage or data. The application of virtualization to storage services or devices for the purpose of aggregating, hiding complexity, or adding new capabilities to lower-level storage

Glossary

resources. Storage can be virtualized simultaneously in multiple layers of a system, for instance to create HSM-like systems.

Storage volume   In CIM, a storage volume is a subclass of CIM_StorageExtent and represents an object presented to an operating system (for example, by a hardware RAID cabinet), to a file system (for example, by a software volume manager), or to another entity. Storage volumes do *not* participate in CIM_StorageRedundancyGroups. They are directly realized in hardware or are the end result of assembling and building on lower-level extents.

Store and forward (switching)   A switching technique that requires buffering an entire frame before a routing decision is made.

Streamed sequence   A new sequence initiated by a sequence initiator in any class of service for an exchange while it already has sequences open for that exchange.

Strip   A number of consecutively addressed blocks in a single extent. A disk array uses strips to map virtual disk block addresses to member disk block addresses. Also known as *stripe element*.

Strip size   Stripe depth.

Stripe   The set of strips at corresponding locations of each member extent of a disk array that uses striped data mapping. The strips in a stripe are associated with each other in a way (for example, relative extent block addresses) that allows membership in the stripe to be quickly and uniquely determined by a computational algorithm. Parity RAID uses stripes to map virtual disk block addresses to member extent block addresses.

Stripe depth   The number of blocks in a strip in a disk array that uses striped data mapping. Also, the number of consecutively addressed virtual disk blocks mapped to consecutively addressed blocks on a single member extent of a disk array.

Stripe element   Strip.

Stripe size   The number of blocks in a stripe. A striped array's stripe size is the stripe depth multiplied by the number of member extents. A parity RAID array's stripe size is the stripe depth multiplied by one less than the number of member extents.

Striped array, striped disk array A disk array with striped data mapping but no redundancy for failure protection. Striped arrays are usually used to improve I/O performance on data that are of low value or easily replaced.

Stripeset Striped array.

Striping Short for data striping; also known as RAID level 0 or RAID 0. A mapping technique in which fixed-size consecutive ranges of virtual disk data addresses are mapped to successive array members in a cyclic pattern; a network technique for aggregating the bandwidth of several links between the same pair of nodes. A single data stream can be spread across the links for higher aggregate bandwidth. Sometimes called *port aggregation*.

Structure of management information (SMI) A notation for setting or retrieving management variables over SNMP.

SSL Secure Sockets Layer.

Subdirectory A directory in a hierarchical directory tree whose parent is a directory.

Subject In the context of access control or authorization, an entity whose access or usage is controlled. Users are examples of subjects.

Substitution The assumption of a component's function in a system by a functionally equivalent component.

SVC Switched virtual circuit.

Swap, swapping The installation of a replacement unit in place of a defective unit in a system. Units are any parts of a system that may be either field replaceable (FRUs) by a vendor service representative or consumer replaceable (CRUs). A physical swap operation may be cold, warm, or hot depending on the state in which the disk subsystem must be in order to perform it. A functional swap operation may be an autoswap, or it may require human intervention.

Switch A network infrastructure component to which multiple nodes attach. Unlike hubs, switches typically have internal band-

Glossary

width that is a multiple of link bandwidth and the ability to rapidly switch node connections from one to another. A typical switch can accommodate several simultaneous full-link-bandwidth transmissions between different pairs of nodes.

Switchback Failback.

Switch-over Fail-over.

Switched over (system) Failed over.

Symmetric cryptosystem A cryptographic algorithm in which the same key is used to encrypt and decrypt a single message or block of stored information. Keys used in a symmetric cryptosystem must be kept secret.

Symmetric virtualization In-band virtualization, which is the preferred term.

Synchronous Digital Hierarchy (SDH) An ISO standard with 155-, 622-, 2048-, and 9953-Mbps serial data rates in steps of 4. A common worldwide telecommunications methodology. SDH uses a light scrambling of data to remove only the lowest-frequency elements with the goal of achieving maximum digital bandwidth use.

Synchronization A receiver's identification of a transmission word boundary; the act of aligning or making two entities be equivalent at a specified point in time.

Synchronous operations Operations that have a fixed time relationship to each other. Most commonly used to denote I/O operations that occur in time sequence; that is, a successor operation does not occur until its predecessor is complete.

Synchronous Optical Network (SONET) Standard for optical network elements. SONET provides modular building blocks, fixed overheads, integrated operations channels, and flexible payload mappings. Basic SONET provides a bandwidth of 51.840 Mbps. This is known as OC-1. Higher bandwidths that are n times the basic rate are available (known as OC-n). OC-3, OC-12, OC-48, and OC-192 are currently in common use.

System board A printed circuit module containing mounting devices for processor(s), memory, and adapter cards and implementing basic computer functions such as memory access, processor and I/O bus clocking, and human interface device attachment.

System disk The disk on which a computer system's operating software is stored. The system disk is usually the disk from which the operating system is bootstrapped (initially loaded into memory). The system disk frequently contains the computer system's swap and/or page files as well. It also may contain libraries of common software shared among several applications.

System under test An entity being tested to verify functional behavior or determine performance characteristics. Distinguished from test system.

T1 copy Mirroring.

T10 The ANSI T10 Technical Committee, the standards organization responsible for SCSI standards for communication between computers and storage subsystems and devices.

T11 The ANSI T11 Technical Committee, the standards organization responsible for Fibre Channel and certain other standards for moving electronic data into and out of computers and intelligent storage subsystems and devices.

Tabular mapping A form of mapping in which a lookup table contains the correspondence between the two address spaces being mapped to each other. If a mapping between two address spaces is tabular, there is no mathematical formula that will convert addresses in one space to addresses in the other.

Tape, tape drive, tape transport A storage device that writes data sequentially in the order in which it is delivered and reads data in the order in which it is stored on the medium. Unlike disks, tapes use implicit data addressing.

Tape array A collection of tapes from one or more commonly accessible storage subsystems, combined with a body of control software.

Tape virtualization, tape drive virtualization, and tape library virtualization The act of creating abstracted tape devices by applying virtualization to tape drives or tape libraries.

Target The system component that receives an SCSI I/O command.

Target ID The SCSI bus address of a target device or controller.

Glossary

TB, Tbyte Abbreviations for terabyte.

TCO Total cost of ownership.

TCP Transmission Control Protocol.

TCP/IP Abbreviation for the suite of protocols that includes TCP, IP, UDP, and ICMP. This is the basic set of communication protocols used on the Internet.

Tenancy The possession of a Fibre Channel arbitrated loop by a device to conduct a transaction.

Terabyte Unit for 1 billion (10^{12}) bytes. SNIA publications typically use the 10^{12} convention commonly found in I/O literature rather than the 1,099,5111,627,776 (2^{40}) convention sometimes used when discussing random access memory.

Test system A collection of equipment used to perform a test. In functional and performance testing, it is generally important to clearly define the test system and distinguish it from the system under test.

Third-party copy (TPC) A protocol for performing tape backups using minimal server resources by copying data directly from the source device (disk or array) to the target device (tape transport) without passing through a server.

Threat Any circumstance or event with the potential to harm an information system through unauthorized access, destruction, disclosure, modification of data, and/or denial of service.

Throughput The number of I/O requests satisfied per unit time. Expressed in I/O requests per second, where a request is an application request to a storage subsystem to perform a read or write operation.

Throughput-intensive (application) A request-intensive application.

Timeserver An intelligent entity in a network that enables all nodes in the network to maintain a common time base within close tolerances.

TNC Threaded Neil Councilman, a type of coaxial cable connector. Specifications for TNC style connectors are defined in MIL-C-39012 and MIL-C-23329.

Token Ring (network) A network in which each node's transmitter is connected to the receiver of the node to its logical right, forming a continuous ring. Nodes on a Token Ring network gain the right to transmit data by retaining a token (a specific unique message) when they receive it. When a node holding the token has transmitted its allotment of data, it forwards the token to the next node in the ring; a LAN protocol for Token Ring networks governed by IEEE Standard 802.3 that operates at speeds of 4 and 16 Mbps.

Topology The logical layout of the components of a computer system or network and their interconnections. Topology deals with questions of what components are connected directly to other components from the standpoint of being able to communicate. It does not deal with questions of physical location of components or interconnecting cables.

Total cost of ownership (TCO) The comprehensive cost of a particular capability such as data processing, storage access, file services, and so on. Total cost of ownership includes acquisition, environment, operations, management, service, upgrade, loss of service, and residual value.

TPC Third-party copy.

Transceiver A transmitter and receiver combined in one package.

Transmission character Any encoded character (valid or invalid) transmitted across a physical interface specified by FC-0. Valid transmission characters are specified by the transmission code and include data characters and special characters.

Transmission code A means of encoding data to enhance their transmission characteristics. The transmission code specified by FC-PH is byte-oriented, with both valid data bytes and special codes encoded into 10-bit transmission characters.

Transmission Control Protocol (TCP) The Internet connection-oriented network transport protocol; TCP provides a reliable delivery service.

Transmission word A string of four contiguous transmission characters aligned on boundaries that are zero modulo 4 from a

Glossary

previously received or transmitted special character. FC-1 transmission and reception operates in transmission word units.

Transmitter The portion of a link control facility that converts valid data bytes and special codes into transmission characters using the rules specified by the transmission code, converting these transmission characters into a bit stream, and transmitting this bit stream on an optical or electrical transmission medium; an electronic circuit that converts an electrical logic signal to a signal suitable for an optical or electrical communications medium.

Transparent fail-over A fail-over from one component of a system to another that is transparent to the external operating environment. Often used to refer to paired disk controllers, one of which exports the other's virtual disks at the same host bus addresses after a failure.

Trap A type of SNMP message used to signal that an event has occurred.

Triaxial cable An electrical transmission medium consisting of three concentric conductors separated by a dielectric material with the spacings and material arranged to give a specified electrical impedance.

Trojan horse A computer program containing hidden code that allows the unauthorized collection, falsification, or destruction of information.

Tunneling A technology that enables one network protocol to send its data via another network protocol's connections. Tunneling works by encapsulating the first network protocol within packets carried by the second protocol. A tunnel also may encapsulate a protocol within itself (for example, an IPsec gateway operates in this fashion, encapsulating IP in IP and inserting additional IPsec information between the two IP headers).

UDP User Datagram Protocol.

ULP Upper-layer protocol.

Ultra SCSI A form of SCSI capable of 20 megatransfers per second. Single-ended Ultra SCSI supports bus lengths of up to 1.5 m. Differential Ultra SCSI supports bus lengths of up to 25 m. Ultra

SCSI specifications define both narrow (8 data bits) and wide (16 data bits) buses. A narrow Ultra SCSI bus transfers data at a maximum of 20 MBps. A wide Ultra SCSI bus transfers data at a maximum of 40 MBps.

Ultra2 SCSI A form of SCSI capable of 40 megatransfers per second. There is no single-ended Ultra2 SCSI specification. Low-voltage differential (LVD) Ultra2 SCSI supports bus lengths of up to 12 m. High-voltage differential Ultra2 SCSI supports bus lengths of up to 25 m. Ultra2 SCSI specifications define both narrow (8 data bits) and wide (16 data bits) buses. A narrow Ultra SCSI bus transfers data at a maximum of 40 Mbps. A wide Ultra2 SCSI bus transfers data at a maximum of 80 MBps.

Ultra3 SCSI A form of SCSI capable of 80 megatransfers per second. There is no single ended Ultra3 SCSI specification. Low-voltage differential (LVD) Ultra2 SCSI supports bus lengths of up to 12 m. There is no high-voltage differential Ultra3 SCSI specification. Ultra3 SCSI specifications only define wide (16 data bits) buses. A wide Ultra3 SCSI bus transfers data at a maximum of 160 MBps.

UML Unified Modeling Language.

Unauthorized disclosure The exposure of information to individuals not authorized to receive or access it.

Unclassified Information that is not designated as classified.

Unicast The transmission of a message to a single receiver. Unicast can be contrasted with broadcast (sending a message to all receivers on a network) and multicast (sending a message to select subset of receivers).

Unicode A standard for a 16-bit character set (each character has a 16-bit number associated with it). Unicode allows for up to 2^{16}, or 65,536, characters, each of which may have a unique representation. It accommodates several non-English characters and symbols and is therefore an aid to development of products with multilingual user interfaces. Unicode was designed and is maintained by the nonprofit industry consortium Unicode, Inc.

Glossary

Unified Modeling Language (UML) A visual approach that uses a variety of diagrams such as use case, class, interaction, state, activity, and others) to specify the objects of a model and their relationships; various tools exist for turning UML diagrams into program code.

Uninterruptible power source (UPS) A source of electrical power that is not affected by outages in a building's external power source; UPSs may generate their own power using gasoline generators, or they may consist of large banks of batteries. UPSs are typically installed to prevent service outages due to external power grid failure in computer applications deemed by their owners to be "mission critical."

Unsolicited control An information category indicated in a Fibre Channel frame header.

Unsolicited data An information category indicated in a Fibre Channel frame header.

Upper-layer protocol (ULP) A protocol used on a Fibre Channel network at or above the FC-4 level; SCSI and IP are examples of ULPs.

UPS Uninterruptible power source.

Usable capacity The storage capacity in a disk or disk array that is available for storing user data. Usable capacity of a disk is the total formatted capacity of a disk minus the capacity reserved for media defect compensation and metadata. Usable capacity of a disk array is the sum of the usable capacities of the array's member disks minus the capacity required for check data and metadata.

User data extent The protected space in one or more contiguously located redundancy group stripes in a single redundancy group. In RAID arrays, collections of user data extents comprise the virtual disks or volume sets presented to the operating environment.

User data extent stripe depth The number of consecutive blocks of protected space in a single user data extent that are mapped to consecutive virtual disk block addresses. In principle, each user data extent that is part of a virtual disk may have a different user data extent stripe depth. User data extent stripe depth may differ from the redundancy group stripe depth of the protected space extent where it resides.

User Datagram Protocol (UDP)   An Internet protocol that provides connectionless datagram delivery service to applications; UDP over IP adds the ability to address multiple end points within a single network node to IP.

User identification number (UID)   A unique number that identifies an individual to a computer system; UIDs are the result of authentication processes that use account names, passwords, and possibly other data to verify that a user is actually who he or she represents himself or herself to be. UIDs are input to authorization processes that grant or deny access to resources based on the identification of the requesting user.

Valid data byte   A string of eight contiguous bits within FC-1 that represents a value between 0 and 255.

Valid frame   A received frame containing a valid start of frame (SOF), a valid end of frame (EOF), valid data characters, and proper cyclic redundancy check (CRC) of the frame header and data field.

Validity control bit   A control bit that indicates whether a field is valid. If a validity control bit indicates that a field is invalid, the value in the field is treated as invalid.

VBA   Virtual block address.

VCI   Virtual channel identifier.

VCSEL   Vertical cavity surface emitting laser.

Vendor unique   Aspects of a standard (for example, functions, codes, and so on) not defined by the standard but explicitly reserved for private usage between parties using the standard. Different implementations of a standard may assign different meanings to vendor-unique aspects of the standard.

Verify, verification   The object-by-object comparison of the contents of a backup image with the online data objects from which it was made.

Glossary

Versioning The maintenance of multiple point-in-time copies of a collection of data. Versioning is used to minimize recovery time by increasing the number of intermediate checkpoints from which an application can be restarted.

Vertical cavity surface emitting laser (VCSEL) A surface-emitting laser source fabricated on a planar wafer with emission perpendicular to the wafer.

VI, VIA Abbreviations for Virtual Interface Architecture.

Virtual block A block in the address space presented by a virtual disk. Virtual blocks are the atomic units in which a virtual disk's storage capacity is typically presented by RAID arrays to their operating environments.

Virtual block address (VBA) The address of a virtual block. Virtual block addresses are typically used in hosts' I/O commands addressed to the virtual disks instantiated by RAID arrays. SCSI disk commands addressed to RAID arrays are actually using virtual block addresses in their logical block address fields.

Virtual channel identifier (VCI) A unique numerical tag contained in an ATM cell header. A VCI identifies an ATM virtual channel over which the cell containing it is to travel.

Virtual circuit Set of state information shared by two communicating nodes that is independent of the particular path taken by any particular transmission; a unidirectional path between two communicating N_Ports. Fibre Channel virtual circuits may be limited to a fraction of the bandwidth available on the physical link.

Virtual device A device presented to an operating environment by control software or by a volume manager. From an application standpoint, a virtual device is equivalent to a physical one. In some implementations, virtual devices may differ from physical ones at the operating system level (for example, booting from a host-based disk array may not be possible).

Virtual disk A set of disk blocks presented to an operating environment as a range of consecutively numbered logical blocks with disklike storage and I/O semantics. The virtual disk is the disk array object that most closely resembles a physical disk from the operating environment's viewpoint.

Virtual Interface Architecture (VIA) An API specification for direct communication among distributed applications developed by Intel, Compaq, and Microsoft; VIA reduces interprocess communication latency by obviating the need for applications to use processor interrupt or operating system paths to intercommunicate while maintaining security on the communications path. VIA is interconnect-neutral.

Virtual path identifier (VPI) An 8-bit field in an ATM cell header that denotes the cell over which the cell should be routed.

Virtual tape A virtual device with the characteristics of a tape.

Virtualization The act of integrating one or more (back end) services or functions with additional (front-end) functionality for the purpose of providing useful abstractions. Typically, virtualization hides some of the back-end complexity or adds or integrates new functionality with existing back-end services. Examples of virtualization are the aggregation of multiple instances of a service into one virtualized service or to add security to an otherwise insecure service. Virtualization can be nested or applied to multiple layers of a system.

Virus A type of programmed threat. A code fragment (not an independent program) that replicates by attaching to another program and either damaging information directly or causing denial of service.

Volatility A property of data. Volatility refers to the certainty that data will be obliterated if certain environmental conditions are not met. For example, data held in DRAM is volatile, since if electrical power to DRAM is cut, the data in it are obliterated.

Volume Virtual disk. Used to denote virtual disks created by volume manager control software. A piece of removable medium that has been prepared for use by a backup manager (for example, by the writing of a media ID).

Volume group A collection of removable media that reside in a single location, for example, in a single robot or group of interconnected robots.

Glossary

Volume manager   Common term for host-based control software.

Volume pool   A logical collection of removable media designated for a given purpose, for example, for holding the copies of a single repetitive backup job or for backing up data from a given client or set of clients. A volume pool is an administrative entity, whereas a volume group is a physical one.

Volume set   Virtual disk.

VPI   Virtual path identifier.

Vulnerability   A weakness in an information system, system security procedures, internal controls, or implementation that could be exploited.

WAN   Wide area network.

Warm spare (disk)   A spare to which power is applied and which is not operating but which otherwise is usable as a hot spare.

Warm swap   The substitution of a replacement unit (RU) in a system for a defective one in which in order to perform the substitution, the system must be stopped (caused to cease performing its function), but power need not be removed. Warm swaps are manual (performed by humans) physical operations.

Wave division multiplexing (WDM)   The splitting of light into a series of "colors" from a few (sparse) to many with a narrow wavelength separation (dense WDM) for the purpose of carrying simultaneous traffic over the same physical fiber (9 μm usually). Each "color" is a separate data stream.

WBEM   Web-based enterprise management.

WDM   Wave division multiplexing; Windows driver model.

Web-based enterprise management (WBEM)   An initiative in the DMTF; it is a set of technologies that enables interoperable management of an enterprise. WBEM consists of CIM, an XML DTD defining the tags (XML encodings) to describe the CIM schema and its data, and a set of HTTP operations for exchanging the XML-based information. CIM joins the XML data description language and HTTP with an underlying information model, CIM, to create a conceptual view of the enterprise.

Well-known address An address identifier used to access a service provided by a Fibre Channel fabric. A well-known address is not subject to zone restrictions; that is, a well-known address is always accessible, irrespective of the current active zone set.

Wide SCSI Any form of SCSI using a 16-bit data path. In a wide SCSI implementation, the data transfer rate in megabytes per second is twice the number of megatransfers per second because each transfer cycle transfers 2 bytes.

Wide area network (WAN) A communications network that is geographically dispersed and that includes telecommunications links.

Windows driver model A Microsoft specification for device drivers to operate in both the Windows NT and Windows 95/98 operating systems.

Windows Internet Naming Service (WINS) A facility of the Windows NT operating system that translates between IP addresses and symbolic names for network nodes and resources.

Windows Management Instrumentation (WMI) The name of the Microsoft framework that supports CIM and WBEM. A set of Windows NT operating system facilities that enables operating system components to provide management information to management agents.

WINS Windows Internet Naming Service.

WMI Windows Management Instrumentation.

Word An addressable unit of data in computer memory. Typically defined to be 16 consecutively addressed bits. Most processor architectures include arithmetic and logical instructions that operate on words. Fibre Channel: The smallest Fibre Channel data element consisting of 40 serial bits representing either a flag (K28.5) plus three encoded data bytes (10 encoded bits each) or four 10-bit encoded data bytes. Fibre Channel: A string of four contiguous bytes occurring on boundaries that are zero modulo four from a specified reference.

Glossary

Workgroup A group of UNIX or Windows computer system users, usually with a common mission or project, that is created for administrative simplicity.

World wide name (WWN) A 64-bit unsigned name identifier that is worldwide unique; a unique 48- or 64-bit number assigned by a recognized naming authority (often via block assignment to a manufacturer) that identifies a connection or a set of connections to the network; a WWN is assigned for the life of a connection (device). Most networking technologies (for example, Ethernet, FDDI, and so on) use a worldwide name convention.

Worm An independent program that replicates from computer to computer across network connections, often-clogging networks and computer systems as it spreads.

Write hole A potential data-corruption problem for parity RAID technology resulting from an array failure while application I/O is outstanding followed by an unrelated member disk failure (some time after the array has been returned to service). Data corruption can occur if member data and parity become inconsistent due to the array failure, resulting in a false regeneration when data from the failed member disk are subsequently requested by an application. Parity RAID implementations typically include mechanisms to eliminate the possibility of write holes.

Write-back cache A caching technique in which the completion of a write request is signaled as soon as the data are in cache, and actual writing to nonvolatile media occurs at a later time. Write-back cache includes an inherent risk that an application will take some action predicated on the write completion signal, and a system failure before the data are written to nonvolatile media will cause media contents to be inconsistent with that subsequent action. For this reason, good write-back cache implementations include mechanisms to preserve cache contents across system failures (including power failures) and to flush the cache at system restart time.

Write penalty Low apparent application write performance to independent access RAID arrays' virtual disks. The write penalty is inherent in independent access RAID data protection techniques, which require multiple member I/O requests for each application write request, and ranges from minimal (mirrored arrays) to substantial (RAID levels 5 and 6). Many RAID array designs include features such as write-back cache specifically to minimize the write penalty.

Write-through cache A caching technique in which the completion of a write request is not signaled until data are safely stored on nonvolatile media. Write performance with a write-through cache is approximately that of a noncached system, but if the data written are also held in cache, subsequent read performance may be improved dramatically.

WWN Worldwide name.

X_ID Exchange identifier.

X3T10 The ANSI X3T10 Technical Committee, the standards organization responsible for SCSI standards for communication between computers and storage subsystems and devices.

X3T11 The ANSI X3T11 Technical Committee, the standards organization responsible for Fibre Channel and certain other standards for moving electronic data in and out of computers.

XML eXtensible Markup Language.

Zero filling The process of filling unused storage locations in an information system with the representation of the character denoting 0.

Zeroization The process of removing or eliminating the key from a cryptographic program or device.

Zone A collection of Fibre Channel N_Ports and/or NL_Ports (that is, device ports) that are permitted to communicate with each other via the fabric. Any two N_Ports and/or NL_Ports that are not members of at least one common zone are not permitted to communicate via the fabric. Zone membership may be specified by port location on a switch (that is, domain ID and port number), the device's N_Port name or the device's address identifier, or the

Glossary

device's node name. Well-known addresses are implicitly included in every zone.

Zone set A set of zone definitions for a fabric. Zones in a zone set may overlap (that is, a port may be a member of more than one zone). Fabric management may support switching between zone sets to enforce different access restrictions (for example, at different times of day).

Zoning A method of subdividing a storage area network into disjoint zones or subsets of nodes on the network. Storage area network nodes outside a zone are invisible to nodes within the zone. Moreover, with switched SANs, traffic within each zone may be physically isolated from traffic outside the zone.

APPENDIX A: QUICK REFERENCE SECTION

How Does a SAN Differ from a LAN or WAN?

A LAN uses network protocols that send smaller amounts of data with increased communication overhead, reducing bandwidth. A SAN uses storage protocols (*Small Computer Systems Interface* [SCSI]), giving it the ability to transmit larger amounts of data with reduced overhead and higher bandwidth.

How Do I Manage a SAN?

Several manufacturers provide SAN management software. Running management software typically requires a separate terminal, such as an NT server, connected to the SAN. Connecting this terminal to a SAN enables additional capabilities, such as zoning, mapping, masking, backup and restore functions, and fault management. An alternative is *Simple Network Management Protocol* (SNMP). SNMP is based on TCP/IP and offers basic alert management, allowing a node to alert the management system of failures of any system component. However, SNMP does not offer proactive management and lacks security.

What Is a SAN Manager?

A SAN Manager is proprietary SAN management software that allows central management of Fibre Channel hosts and storage devices. A SAN Manager enables systems to use a common pool of storage devices on a SAN, enabling SAN administrators to take full advantage of storage assets and reduce costs by leveraging existing equipment more efficiently.

Can My Legacy System Be Integrated into a SAN?

It is possible to integrate most legacy SCSI storage systems and servers into a SAN environment. This allows you to leverage your existing investments because you can locate your storage miles away as opposed to only a few feet.

Should I Use a Switch or a Hub?

Switches provide several advantages in a SAN environment. If a single switch fails in a switched-fabric environment, other switches in the fabric remain operational, whereas a Hub-based environment generally fails if a single hub on the loop fails. Further, switches support the *Fibre Channel Switch* (FC-SW) standard, making addressing independent of the subsystem's location on the fabric, and provide superior fault isolation along with high availability. FC-SW also allows the host to better identify subsystems connected to the switch. In terms of scalability, interconnection switches provide thousands of connections without degrading bandwidth, whereas a hub-based loop is limited to 126 devices. In terms of availability, switches support the online addition of subsystems (servers or storage) without requiring reinitialization or shutdown. Hubs require a *loop initialization* (LIP) to reacquire subsystem addresses every time a change occurs on the loop. A LIP typically takes 0.5 s and can disable a tape system during the backup process.

Appendix A: Quick Reference Section

What Is Fibre Channel?

Fibre Channel is the enabling technology behind SANs. An industry standard storage/networking interface, it connects host systems, desktop workstations, and storage devices by means of a point-to-point bidirectional serial interface. Fibre Channel is capable of transmitting data at high speeds with low latency over long distances—at 1 Gb speed (200 MBps full duplex) and at distances of up to 10 km. Fibre Channel is a transport mechanism for supported protocols (ATM, FDDI, TCP/IP, HIPPI, SCSI, and so on) that combines the connectivity and distance of the networking protocol with the simplicity and reliability of channel switching via the same physical wire (either copper or fiberoptic medium). The interface used to connect Fibre Channel cabling to hosts and storage devices is called a *host bus adapter* (HBA). Each port uses a pair of fibers for two-way communications, with the *transmitter* (TX) connecting to a *receiver* (RX) at the other end of the Fibre Channel cable.

What Is Fibre Channel Fabric?

A *fabric* is at least one Fibre Channel switch in a networked topology. Fabric is a routing technology that attaches devices to itself in point-to-point fashion. Fabric establishes high-bandwidth connections between nodes on the the fabric by using unique identifiers.

How Do SCSI Tape Drives Connect to a Fibre Channel SAN?

Because nearly every SAN implementation uses Fibre Channel technology, an industry standard network interface is necessary. Fibre Channel connectivity requires that an HBA be attached to every server and storage device on the SAN. Each port uses a pair of fibers for two-way communications, with the TX connecting to a RX at the other end of the Fibre Channel cable.

What Is an Interconnect?

An *interconnect* is what connects the different components of a SAN. It is used for high-speed, high-bandwidth connections within a SAN. It can access data at speeds 100 times faster than current networks and connects all the pieces of a SAN, providing scalability, connectivity, performance, and availability. *Input/output* (I/O) buses and networks are examples of interconnects.

What Five Questions Should You Ask Prior to Building a SAN?

- What problems are being resolved through implementation of the SAN environment?
- What is the technical requirement for the business operation?
- What is the broad case of the business environment?
- What are the corporate and/or project goals in SAN implementation?
- Is there a timeline, and is there a cost and benefit analysis that needs to be recognized?

Who Are the Major SAN Vendors?

The major SAN vendors include Hewlett-Packard, Compaq, ITIS, McData, StorageTek, Hitachi Data Systems, Brocade, Gadzoox, Veritas, Legato, Computer Associates, Oracle, DataCore, Ancor, Vixel, Sun, Dell, and others.

INDEX

Symbols

10 Gb Ethernet, 67, 71–72
10 Gb Fibre Channel, 67
12/2/160 architecture, 21
16/2/224 architecture, 20
16/4/192 architecture, 19
8/2/96 architecture, 18
8B/10B transmission characters, Fibre Channel, 110

A

A.B. Watley Group case study, 143, 147
    SAN planning, 146
    Sun hardware solutions, 144
ABR, ATM QoS, 74
access, 151
access control, 151
access control list, 151
access fairness, 151
access method, 151
access path, 151
ACKs, Fibre Channel, 81
ACL, 151
ACS, 151
active, 151–152
active component, 152
active copper, 151
adapter, 152
adapter card, 152
adaptive array, 152
address, 152
Address identifier, 81, 152
address resolution, 152
Address Resolution Protocol, 152
addressing, 108, 153
administration host, 153
administrator, 153
Advanced Encryption Standard, 153
Advanced Features layer, FC-3, 13, 94
advanced intelligent tape, 153
AES, 153
agent, 153
aggregation, 153
AH, 153
AIT, 153
algorithmic mapping, 153
alias, 154
alias address identifier, 154
alternate client restore, 154
alternate path restore, 154
always on, 154
AL_PA (arbitrated-loop physical address), Fibre Channel, 81, 101
AL_TIME (arbitrated-loop timeout value), Fibre Channel, 81
American National Standards Institute (ANSI), 154
ANSI T10, 154
ANSI T11, 154
ANSI X3T10, 155
ANSI X3T11, 155
API, 155
appliance, 155
application I/O request, 155
application programming interface, 155
Application Response Measurement, 155
application write request, 155
application-specific integrated circuit, 155
ARB (arbitrate primitive signal), Fibre Channel, 82

Arbitrated Loop, Fibre Channel, 15, 81, 101–102, 155
   physical address, 156
arbitration, 156
ARBx, Fibre Channel, 101
architecture, Fibre Channel, 17, 91
   12/2/160, 21
   16/2/224, 20
   16/4/192, 19
   8/2/96, 18
architecture of SANs, 40
archive/archiving, 156
ARM, 156
ARP, 156
array, 156
ASIC, 156
ASPs (Application Service Providers), case study example, 129
Association_Header, 156
asymmetric cryptosystem, 156
asymmetric virtualization, 157
asynchronous I/O operation, 157
asynchronous I/O request, 157
Asynchronous Transfer Mode, 157
AT&T, IP-enabled Frame Relay, 75
ATM, 157
   cells, 73
   overhead, 74
   PVCs, 73
   QoS, 74
   SVCs, 74
   WAN data traffic, 70–71
atomic operation, 157
attenuation, 157
audit trail, 158
authentication, 158
authentication header, 158
authorization, 158
auto swap, 158

automated cartridge system, 158
automatic backup, 158
automatic fail-over, 158
automatic swap, 158
automatic switchover, 158
availability, 158

B

backing store, 159
backup client, 159
backup image, 159
backup manager, 159
backup policy, 159
backup window, 159
backups
   centralized, 35
   clients, 37
   LAN-free, 35
bandwidth, 159
bandwidth crisis, 59–60
basic input-output system, 159
baud, 160
Bayonet Neil Councilman (BNC) connector, 160
BB_Buffer, 160
BB_Credit (buffer-to-buffer credit value), Fibre Channel, 82, 160
beginning running disparity, 160
benefits of SANs, 2–4
BER, 160
Berkeley RAID levels, 160
best effort class of service, 160
Big Endian, 160
BIOS, 160
bit error rate, 160
bit synchronization, 161
black, 161

Index

blind mating, 161
block, 161
block addressing, 161
block virtualization, 161
BNC, 161
Boise Cascade Corp. case study, 122
    backup time savings, 124
    home directory cleanup campaign, 123
boot, 161
booting, 161
bootstrapping, 161
bridge controller, 162
broadcast, 162
buffer, 162
buffer-to-buffer flow control, 162
Bus-&-Tags, SAN interfaces, 3
Butler, Michael, SANs and business continuity, 126–127
bypass circuit, 162
Byte, 162
B_Port, 159

C

C-H-S addressing, 164
CA, 162
cable plant, 162
cache, 163
canister, 163
carousel, 163
Carrier Sense Multiple Access with Collision Detection, 163
cascading, 163
case studies
    A.B. Watley Group, 143
        SAN planning, 146
        Sun hardware solutions, 144
    Boise Cascade Corp., 122
        backup time savings, 124
        home directory cleanup campaign, 123
    DTCC, 139
        backup speeds, 141
        high support level, 142
        redundancy assurance, 139–140
        safety and soundness issues, 140–141
    Euroconex
        disaster recovery, 138
        high-availability data storage, 136–137
    New York Life, 116
        analyzing storage needs, 119
        cost savings, 118
        independent consultant value, 119
        SAN environment complexity, 118
        SAN implementation, 117
    Orchid BioSciences
        data restoration, 135
        virus attacks, 133–134
    R&B Group, 120–121
    TDS Informationstechnologie AG, 129
        benefits of solution, 132
        criteria for solution selection, 131
        high-availability storage, 130
        storage on demand, 131
catalog, 163
CBR, ATM QoS, 74
CDR, 163
cells, ATM, 73
centralized backups, 35
certification authority, 164
changed block-changed block point-in-time copy, 164
channels, 11, 80, 164
character, 164
character cell interface, 164
check data, 164
checkpoint, 164

chunk, 164
chunk size, 164
CIFS, 165
CIM, 165
cipher, 165
ciphertext, 165
circuit, 165
CKD architecture, 165
Class 1 service, Fibre Channel, 82, 97, 165
Class 2 service, Fibre Channel, 82, 97, 165
Class 3 service, Fibre Channel, 82, 98, 165
Class 4 service, Fibre Channel, 99, 165
Class 5 service, Fibre Channel, 99
Class 6 service, Fibre Channel, 100, 165
class of service, 166
classified information, 165
cleartext, 166
CLI, 166
client service request, 166
clients, 166
　backup systems, 37
CLS (close primitive signal), Fibre Channel, 82
cluster, 166
CMIP, 166
coaxial cable, 166
code balance, 166
code bit, 166
code byte, 167
code violation, 167
cold backup, 167
cold swap, 167
combined switched access and shared media SANs, 26–27
comma character, 167
command line interface, 167
Common Information Model, 167
Common Internet File System, 167
Common Management Information Protocol, 167

communication circuit, 168
communications security, 168
complex array, 168
components of SANs, 3–4
compression, 34, 168
computer security, 168
concatenation, 168
concurrency, 168
concurrent, 169
concurrent copy, 169
concurrent operations, 169
conditioning, 169
confidentiality, 169
configuration, 169
configuring SANs
　architecture, 40
　from legacy systems
　　four-to-one modeling, 38
　　LAN-free backups, 35
　　tape drives, 35
　system requirements analysis, 38
connection, 169
connection initiator, 169
connection recipient, 169
connectionless buffer, 169
connectionless frame, 170
connectionless integrity service, 170
connectionless service, 170
console, 170
consolidation, 170
continuity of business, SANs, 126–127
continuously increasing relative offset, 170
control software, 170
controller, 170
controller cache, 171
controller-based array, 171
controller-based disk array, 171
copper cable, Fibre Channel, 104
copy on write, 171
copyback, 171

Index

cost advantages of DWDM, 64–66
cost savings, New York Life case study, 118
count-key-data, 171
covert channel, 171
CRC, 172
credit, 172
CRU, 172
cryptanalysis, 172
cryptosystem, 172
CSMA/CD, 172
cumulative incremental backup, 172
current running disparity, 172
customer delivery of legacy-to-SAN storage solutions, 43
customer-replaceable unit, 172
cut through switching, 172
cyclic redundancy check, 172
cylinder-head-sector addressing, 173

D

daemon, 173
DAFS (Direct Access File System), 29
DAS (direct-attached storage), 56
DAS (directly attached storage), 29
data availability, 173
Data byte, 173
data character, 173
Data Encryption Standard, 177
Data frame, 173
data manager, 173
data model, 173
data reliability, 174
data stripe depth, 174
data striping, 174
data traffic
  bandwidth crisis, 59–60
  SONET issues, 70
  WANs, 69–71
data transfer capacity, 174
data transfer rate, 174
data transfer-intensive application, 174
database management system, 174
datagram, 175
Datagram service, Fibre Channel, 98
DBMS, 175
decoding, 175
decryption, 175
dedicated connection, 175
dedicated connection service, 175
degaussing, 175
delimiter, 175
DEN, 175
denial of service, 175
DES, 175
designing SANs, goals of design, 5
Desktop Management Interface, 175
Destination identifier, 175
Destination N_Port, 176
development of SAN architecture.
  See SAN configuration.
device, 176
device bus, 176
device channel, 176
device discovery, SNS, 26
device fanout, 176
DHCP, 176
differential signaling, 176
differential incremental backup, 176
Differentiated Services (DiffServ), 176–177
digest, 177
digital linear tape, 177
digital signature, 177
directory, 177
Directory Enabled Network, 177
directory tree, 177
disaster tolerance, Fibre Channel SANs, 28
discard policy, 177
disconnection, 178

disk array, 178
disk array subsystem, 178
disk block, 178
disk cache, 178
disk drive, 178
disk enclosures, Fibre Channel, 104
disk image backup, 178
disk scrubbing, 179
disk shadowing, 178
disk striping, 178
disk subsystem, 179
disks, 178
   Fibre Channel, 105
disparity, Fibre Channel, 83, 179
disparity control, 93
Distributed Management Task Force, 179
DLT, 179
DMI, 179
DMTF, 179
DNS, 179
document type definition, 179
domain, 179
domain controller, 179
Domain Name Service, 179
DoS, 179
double buffering, 180
drawbacks of SAN implementation, 42
drive letter, 180
drive sharing, 37
drivers, 105, 180
DSA, 180
DTCC case study (Depository Trust & Clearing Corporation), 139
   backup speeds, 141
   high support level, 142
   redundancy asurance, 139–140
   safety and soundness issues, 140–141
DTD, 180
dual active components, 180
duplicate, 180

DWDM
   bandwidth crisis solution, 60
   cost reductions, 64–66
   expandability, 63
   flexibility, 62
   optical amplifiers, 62
   overview of technology, 61
   SAN transition model, 58
Dynamic Host Control Protocol, 180
dynamic mapping, 180
D_ID field, Fibre Channel, 82, 173

E

EBU, 181
ECC, 181
EDFAs (erbium-doped fiber amplifiers), DWDM, 62
EE_buffer, 181
EE_Credit (end-to-end credit value), 83, 181
electronic storage element, 181
embedded controller, 181
encapsulating security payload, 181
encoding, 181
encryption, 181
end of frame, 181
end-to-end encryption, 181
end-to-end flow control, 181
enterprise management console, 170
Enterprise Resource Management, 181
Enterprise Systems Connection, 182
Entry port-exit port, 182
EOF (end-of-frame delimiter), Fibre Channel, 83, 182
ERM, 182
Error-correcting code, 182
ESCON interfaces (Enterprise Systems Connection), 3, 23, 53, 83, 182
ESP, 182

Index

ESRM, 182
Ethernet, 182
  10 Gb, 67, 71–72
  adapter, 183
  MANs, 68–69
Euroconex case study
  disaster recovery, 138
  high-availability data storage, 136–137
European Broadcast Union, 183
EVaults, FMC case study, 127–129
EVSN, 183
Exabytes, 2
examples of SAN transitions, 56
  DWDM, 58
  peer models, 57
exchange, Fibre Channel, 83, 112
exchange identifier, 183
exchange status block, 183
exclusive connection, 183
Exit port, 183
expandability of DWDM, 63
expansion card, 183
expansion module, 183
expansion slot, 184
explicit addressing, 184
export, 184
extenders, Fibre Channel, 105
eXtensible Markup Language, 184
extent, 184
external controller, 184
external disk controller, 184
external storage controller, 184
external volume serial number, 184
eye, 184
eye diagram, 185
eye opening, 185
E_D_TOV (error detect timeout value),
    Fibre Channel, 83
E_Port, Fibre Channel, 96, 181

F

F/NL_Port, Fibre Channel, 84
Fabric login, 185
  Fibre Channel, 109
  NICs, 26
fabric name, 185
Fabric Switched Fibre Channel topology, 16
fabrics, 8, 185
  Fibre Channel, 83, 104
    switches, 8
  SANs, 4
fail-over, 190
failback, 185
failed over, 185
failure tolerance, 186
fanout, 186
Fast Ethernet, 68
Fast SCSI, 186
fault tolerance, 186
FBA, 186
FC-0, 13, 84, 91–92, 186
FC-1, 13, 84, 91–92, 186
FC-2, 13, 84, 91–94, 186
FC-3, 13, 84, 91, 94, 187
FC-4, 14, 84, 92, 95, 187
FC-AE, 187
FC-AL (Fibre Channel Arbitrated Loop),
    15–16, 28, 84, 187
FC-AV, 187
FC-GS2, 187
FC-PH (Fibre Channel physical standard),
    84, 92, 186
FC-SB, 187
FC-SB2, 187
FC-SW, 187
FC-SW2, 187
FC-VI, 187
FCA, 187

FCIA (Fibre Channel Industry
    Association), 29
FCIP (Fibre Channel over IP), 85
FCPs (Fibre Channel protocols), 11, 187
FCSI, 187
FDDI, 187
FDDI adapter, 187
Federal Information Processing
    Standard, 187
Federated Management Architecture
    Specification, 187
fiber, 188
Fiber Distributed Data Interface, 188
fiber-optic cable connectors, 85, 105
Fibre Channel, 188
    10 Gb, 67
    8B/10B transmission characters, 110
    addressing, 108
    arbitrated loop, 101–102
    ARBx, 101
    architecture, 91
    channels, 11, 80
    copper cables, 104
    disk enclosures, 104
    disks, 105
    drivers, 105
    exchanges, 112
    extenders, 105
    fabric login, 109
    Fabric Switched topology, 16
    fabric topology, 104
    FC-0, 92
    FC-1, 92
    FC-2, 93–94
    FC-3, 94
    FC-4, 95
    FC-AL topology, 15–16
    fiber-optic cable connectors, 105
    FL_Ports, 9
    four-to-one tape drive modeling, 38
    frames, 111
    F_Ports, 9
    GBICs (Gigabit interface converters), 105
    GLMs (Gigabit link modules), 106
    HBAs, 49, 106
    hubs, 106
    initiators, 9
    interfaces, 80
    IWU (interworking unit), 108
    layers, 13
    LCF, 9, 94
    legacy-to-SAN environment, 42
    LILP, 103
    link analyzers, 107
    LIRP, 103
    loop initialization, 102–103
    multimode cable, 107
    network topologies, 8
    networks, 80
    NL_Ports, 8–9
    N_Ports, 8
        identifiers, 9
        login, 109
    overview of operation, 9
    PBC (port bypass circuit), 106
    point-to-point topology, 15, 100
    ports, 95
    routers, 107
    SAN interface, 3
    SCSI switches, 107
    sequences, 112
    service classes, 96–99
    static switches, 107
    switch WAN extender, 108
    targets, 9
    terminology, 81–91
    topologies, 100
    transmission characters, 110
    transmission words, 111
    underpinning of SAN, 6

Index

Fibre Channel Arbitrated Loop, 188
Fibre Channel architecture, 17
   12/2/160, 21
   16/2/224, 20
   16/4/192, 19
   8/2/96, 18
Fibre Channel Association, 188
Fibre Channel Avionics Environment, 188
Fibre Channel Community, 188
Fibre Channel Generic Services, 189
Fibre Channel Industry Association, 189
Fibre Channel Loop Community, 188
Fibre Channel name, 189
Fibre Channel Protocol, 189
Fibre Channel SANs
   combined switched access and shared media, 26–27
   disaster tolerance, 28
   JBODs, 26
   loop tenancy, 28
   management facilities, 27
   management tools, 28
   NICs, fabric login, 26
   RAID, 26
   shared-media SANs, 26
   SNS, device discovery, 26
   switched-access networks, 26
Fibre Channel Service Protocol, 189
Fibre Channel Single Byte Command Set, 189
Fibre Channel Switched Fabric Interconnect, 189
Fibre Channel Systems Initiative, 189
Fibre Channel Virtual Interface, 189
Fibre Connect, 190
FICON (Fibre Connection), 54, 84, 190
field-replaceable unit, 190
file, 190
file server, 190
file system, 190

file system virtualization, 190
file virtualization, 190
filer, 191
Fill byte, 191
Fill word, Fibre Channel, 84
FIPS, 191
firmware, 191
fixed block architecture, 191
flexibility of DWDM, 62
FLOGI, 191
fluoride-based optical amplifiers, 62
FL_Ports (fabric-loop port), Fibre Channel, 9, 84, 96, 191
FMC case study (Fraser Millner Casgrain), EVaults implementation, 127–129
formatting, 191
four-to-one modeling
   SAN configuration, 38
   shared libraries, 37
Fox, Gary, nationwide SAN implementation, 125–126
frames, 191
   content, 192
   Fibre Channel, 85, 111
Framing and Signaling layer, Fibre Channel FC-2, 13, 93–94
FRU, 192
FSP, 192
full backup, 192
full duplex, 192
full volume transfer rate, 192
functional requirements of SANs, 4
F_BSY (fabric port busy frame), Fibre Channel, 83
F_Port name, 185
F_Ports (fabric port), Fibre Channel, 9, 85, 96, 185
F_RJT (fabric port reject frame), Fibre Channel, 85

G

gateways, SAN-based, 53–54
GB, 192
GBE, 192
GBICs (Gigabit interface converters), Fibre Channel, 105, 193
Gbit, 192
Gbyte, 192
geometry, 193
gigabaud link module, 193
Gigabit, 192–193
Gigabit Ethernet, 68, 193
gigabit interface converter, 193
Gigabits, 34
Gigabyte, 192–193
Gigabyte System Network, 193
GLMs (Gigabit link modules), Fibre Channel, 106, 193
graphic user interface, 194
group, 194
GSN, 194
GUI, 194
G_Port, Fibre Channel, 96, 192

H

hacker, 194
Hard zone, 194
HBAs (host bus adapters), Fibre Channel, 3, 11, 29, 49, 106, 194
headers, ATM cells, 73
hierarchical storage management, 194
high availability, 194
High-Performance Parallel Interface, 194
high-speed serial direct connect, 195
HIPPIs (High-Performance Parallel Interfaces), 3, 11, 23, 80, 195
host, 195
host adapter, 195
host bus adapter, 195
host cache, 195
host computer, 195
host environment, 196
host I/O bus, 196
host-based array, 195
host-based virtualization, 195
hot backup, 196
hot disk, 196
hot file, 196
hot spare, 196
hot standby, 196
hot swap, 196
hot-swap adapter, 196
HSM, 197
HSSDC, 197
HTML, 197
HTTP, Fibre Channel SAN management, 28, 197
hub port, 197
hubs, 106, 197
hunt groups, 94, 197
HyperText Markup Language, 197
HyperText Transfer Protocol, 197

I

I/O, 201
I/O adapter, 202
I/O bottleneck, 202
I/O bus, 202
I/O driver, 202
I/O intensity, 202
I/O load, 203
I/O operation, 203
I/O request, 203
I/O subsystem, 203
ICMP, 197
idempotency, 198

Index

Idle, Fibre Channel, 85
Idle word, 198
IETF, 198
Ignored field, 198
IKE, 198
implicit addressing, 198
importance of data storage, 2
in-band virtualization, 198
INCITS (International Committee for Information Technology Standards), 81
incremental backup, 198
  shared libraries, 37
independent access array, 198
independent consultants, New York Life case study, 119
Infiniband, 23, 67
infinite buffer, 198
information category, 199
information model, 199
information system, 199
information technology, 199
information unit, 199
infrastructure-based virtualization, 199
inherent cost, 199
initial relative offset, 199
initialization, 200
initiator, 200
  Fibre Channel, 9
instantiation, 204
intelligent controller, 204
intelligent device, 204
intelligent peripheral interface, 204
intercabinet, 204
interconnects, 204
  SANs, 3, 11
    Fibre Channel switches, 8
  schemes. *See* fabrics.
interfaces
  connectors, 200
  Fibre Channel, 80
  SANs, 3

intermix, 200
  Fibre Channel, 86, 100
International Standards Organization, 200
Internet Control Message Protocol, 201
Internet Engineering Task Force, 201
Internet Key Exchange, 201
Internet Protocol, 201
interoperability issues, SAN development
  gateways, 53–54
  mainframe-open system data access, 52
  placement, 54
interrupt, 201
interrupt switch, 201
intracabinet, 201
ION (Integrated On-Demand Network), Sprint, 75
IP, 203
IP Security, 203
IP-enabled Frame Relay, AT&T, 75
IPI, 203
IRR (internal rate of return), 50
ISCSI (Internet SCSI), 22
  legacy-to-SAN environment, 43
  SANs, 23
ISO, 203
IT, 203
IWU (interworking unit), Fibre Channel, 108
I_Node, 200

J–K

Java, 203
JBODs (just a bunch of disks), 2, 29, 203
  combined switched access and shared media SANs, 26
Jini, 204
Jiro, 204
jitter, 204
K28.5, 204

KB, 204
Kbyte, 204
key, 204
key exchange, 204
key management, 204
key pair, 204
keying material, 204
Kilobyte, 205

L

label, 205
LANs (Local Area Networks), 205
   data storage, 35
   library sharing, four-to-one modeling, 37
   tape libraries, 37
LAN switches, Fibre Channel, 107
LAN-free backups, 35, 205
LANE, 205
large I/O request, 205
latency, 205
latent fault, 205
layers, Fibre Channel, 13
LBA, 205
LCF (link control facility), Fibre Channel, 9, 94
LDAP, 205
LDM, 205
LED, 205
legacy system SAN configuration
   four-to-one modeling, 38
   LAN-free backups, 35
   tape drives, 35
legacy-to-SAN environment
   customer delivery of storage solution, 43
   drawbacks, 42
   Fibre Channel communication infrastructure, 41–42

interoperability issues
   mainframe-open system data access, 52
   SAN gateways, 53–54
   SAN placement, 54
iSCSI/Fibre Channel, 43
NAS (network area storage), 42
ROI, 47–50
set protocol solutions, 42
storage architecture and design, 45–46
storage performance surveys, 44
libraries, 206
library sharing, LAN data storage, 37
LIFA (loop initialization fabric assigned frame), Fibre Channel, 86
life cycles of SANs, 39
light-emitting diode, 206
Lightweight Directory Access Protocol, 206
LIHA (loop initialization hard assigned frame), Fibre Channel, 86
LILP (loop initialization loop position frame), Fibre Channel, 86, 103
link, 206
link analyzers, Fibre Channel, 107
link services, Fibre Channel, 86
link switches, Fibre Channel, 107
LIP (loop initialization process), 27, 86, 206
LIPA (loop initialization previously assigned frame), Fibre Channel, 86
LIRP (loop initialization report position frame), Fibre Channel, 87, 103
LISA (loop initialization soft assigned frame), Fibre Channel, 87
LISM (loop initialization select master frame), Fibre Channel, 87, 206
load balancing, 206
load optimization, 206
load sharing, 207
local area network, 207
   emulation, 207

Index

Local F_Port, 207
logical block, 207
logical block address, 207
logical disk, 207
logical disk manger, 208
logical unit, 208
logical unit number, 208
logical volume, 208
login, Fibre Channel, 109
login server, 208
long-wavelength laser, 208
loop initialization, 102–103, 208
loop initialization primitive, 208
loop initialization select master, 208
loop port state machine, 209
loop tenancy, Fibre Channel SANs, 28
loopback, 208
LPB (loop port bypass primitive sequence), Fibre Channel, 87
LPE (loop port enable primitive sequence), Fibre Channel, 87
LPSM (loop port state machine), Fibre Channel, 87
LR (link reset primitive sequence), Fibre Channel, 88
LRR (link reset response primitive sequence), Fibre Channel, 88
LUN, 209
LWL, 209
L_Port, Fibre Channel, 87, 96, 205

M

MAC, 209
magnetic remanance, 209
mainframe-open system data access, 52
Managed Object Format, 209
management facilities, Fibre Channel SANs, 27
Management Information Base, 209
management tools, Fibre Channel SANs, 28
managing SAN project, 55
mandatory provision, 209
MANs, 209
  10 Gb Ethernet, 67
  Ethernet, 68–69
mapping, 209
mapping boundary, 209
maximum transfer unit, 209
MB, 209–210
Mbit, 210
MBps, 210
Mbyte, 209
MD5, 210
mean time between failures, 210
mean time to (loss of) data availability, 210
mean time to data loss, 210
mean time to repair, 210
Meaningful control field, 210
media access control, 211
Media ID, 211
media manager, 211
media robot, 211
media stacker, 211
medium, 211
Megabaud, 211
Megabit, 211
Megabyte, 211
Megatransfer, 211
member, 211
member disk, 211
message-digest algorithm, 211
metadata, 212
metropolitan area network, 212
MIB, 212
MIME, 212

mirrored array, 212
mirrored disks, 212
mirroring, 212
MLS, 212
modeling language, 212
MOF, 212
monitor program, 212
mount, 213
MPoA (Multiprotocol over ATM), 75
MRK (mark primitive signal), Fibre Channel, 88
MTBF, 213
MTDA, 213
MTDL, 213
MTTR, 213
MTU, 213
multicast, 213
   Fibre Channel, 94
multicast group, 213
multilevel disk array, 213
multilevel security, 213
multimode fiberoptic cable, 107, 213
multipath I/O, 214
multiplex, Fibre Channel, 97
Multipurpose Internet Mail Extensions, 214
multithreaded, 213

N

NAA, 214
Name identifier, 214
name server, 214
namespace, 214
naming, 214
NAS (network-attached storage), 30-31, 42, 56, 215
National Committee on Information Technology (IT) Standards, 215
National Institute of Standards and Technology, 215
nationwide SANs, 125–126
NCITS, 215
NDMP, 215
network adapter, 215
network address authority, 215
network-attached storage, 216
Network Data Management Protocol, 216
Network File System, 216
network interface card, 216
networks, 215
   Fibre Channel, 80
      topologies, 8
New York Life case study, 116
   analyzing storage needs, 119
   cost savings, 118
   independent consultant value, 119
   SANs
      environment complexity, 118
      implementation, 117
NFS, 216
NICs, 217
   fabric login, 26
NIST, 217
NL_Ports, Fibre Channel, 8–9, 88, 96, 217
node, 217
node name, 217
non-OFC laser, 217
nonlinear mapping, 217
Nonparticipating mode, Fibre Channel, 88
nonrepeating ordered set, 217
nonrepudiation, 217
nontransparent fail-over, 218
nonuniform memory architecture, 218
nonvolatile random access memory, 218
nonvolatility, 218
Normal mode, 217
normal operation, 217

Index

NOS (not operational primitive sequence), Fibre Channel, 88
not operational, 218
NPV (net present value), 50
NUMA, 218
NVRAM, 218
NVRAM cache, 218
NVRAM card, 219
N_Ports, Fibre Channel, 8, 95
   identifiers, 9
   login, 109
   names, 89, 214

O

object, 219
object-oriented methodology, 219
OC-n, 219
OFC, 219
offline backup, 219
OLS (offline primitive sequence), Fibre Channel, 89
online backup, 219
   EVaults, 128
OO, 219
open, 219
open fiber control, 219
Open Group, 220
open interconnect, 220
open system-mainframe data access, 52
operating environment, 220
operation, 220
operation associator, 220
operation of Fibre Channel, 9
operational state, 220
OPN (open primitive signal), Fibre Channel, 89
optical amplifiers, DWDM, 62

optical fall time, 220
optional characteristic, 220
Orchid BioSciences case study
   data restoration, 135
   virus attacks, 133–134
ordered set, 89, 220
organizations connected to SAN technology, 29
originator, 89, 221
originator exchange identifier, 221
out-of-band transmission, 221
out-of-band virtualization, 221
overhead, ATM, 74
overview s
   DWDM, 61
   Fibre Channel operation, 9
Overwrite procedure, 221
OX_ID (originator exchange identifier), Fibre Channel, 89, 221

P

panic, 221
parallel transmission, 221
parallel access array, 221
parity disk, 222
Parity RAID, 222
Parity RAID array, 222
Participating mode, Fibre Channel, 89
partitioning, 222
passive, 156
passive copper, 222
passphrase, 222
password, 222
path, 222
path length, 223
path name, 223
payback period, 50

payload, 223
PB, 223
PBC (port bypass circuit), Fibre Channel, 106, 223
Pbyte, 223
PCI, 223
PCNFSD, 223
PDC, 223
peer model, 57
penetration, 223
peripheral component interconnect, 223
persistence, 223
physical block, 223
physical block address, 224
physical configuration, 223
physical disk, 224
physical extent, 224
physical extent block number, 224
Physical layer, Fibre Channel FC-0, 13, 92
PKI, 224
plain text, 224
PLDA, 224
PLOGI, 224
point-in-time copy, 224
point-to-point Fibre Channel topology/connections, 8, 15, 100
pointer copy, 224
pointer remapping, 225
policy processor, 225
port bypass circuit, 225
port login, 225
port name, 225
ports, Fibre Channel, 95, 225
Port_ID, 225
POST, 225
power conditioning, 225
power-on self-test, 226
present, 226
primary domain controller, 226
primitive sequences, Fibre Channel, 89, 226

primitive signals, Fibre Channel, 90, 226
private key, 226
private key cryptography, 226
private loop
   device, 226
   Fibre Channel, 90
Private NL_Port, Fibre Channel, 90
process associator, 226
process policy, 226
profile, 227
project management, 55
Project Pronto, SBC, 75
proprietary I/O bus, 227
proprietary interconnect, 227
protected space, 227
protected space extent, 227
Protocol Mapping layer, Fibre Channel FC-4, 14, 95
protocols, 227
   DAFS, 29
   Fibre Channel, 11
Psec, 203
public key, 227
public key cryptography, 227
public key infrastructure, 227
public loop, 90, 227–228
Public NL_Port, Fibre Channel, 90
pull technology, 228
push technology, 228
PVCs (permanent virtual circuits), ATM, 73, 228

Q

QoS (Quality of Service), 228
   ATM, 74
   WANs, 70
quiesce, 228
Quiescent state, 228

Index

R

R&B Group case study, 120–121
RAID (redundant array of independent disks), 2, 30, 228
  combined switched access and shared media SANs, 26
RAID 0, 229
RAID 2, 229
RAID 3, 229
RAID 4, 229
RAID 5, 229
RAID 6, 229
RAID Advisory Board, 229
RAM disk, 229
random I/O, 229
random relative offset, 230
rank, 230
RAS, 230
raw partition, 230
raw partition backup, 230
read/write head, 230
real-time copy, 230
rebuilding, 230
receiver, 231
receptacle, 231
reconstruction, 231
recorded volume serial number, 231
recovery, 231
red, 231
red/black concept, 231
Reduced mode, 231
reduction, 231
redundancy, 231–232
  Fibre Channel architecture, 17
redundancy group, 232
redundancy group stripe, 232
redundancy group stripe depth, 232
redundant array of independent disks, 233
regeneration, 233

registered state change notification, 233
rekeying, 233
relative offset, 233
relative offset space, 233
removable media, 233
removable media storage device, 233
repeater, 233
repeating ordered set, 234
replacement disk, 234
replacement unit, 234
replay attack, 234
replica, 234
replicate, 234
Request For Comment, 234
request-intensive application, 234
requirements analysis, SAN configuration, 38
Reserved field, 235
resilience, Fibre Channel architecture, 17
responder, Fibre Channel, 90, 235
responder exchange identifier, 235
restoration, 235
retention period, 235
retimer, 235
return loss, 235
RFC, 235
robot, 235
robotic media handler, 235
ROI, SAN development, 47–50
rotational latency, 236
routers, Fibre Channel, 107
row, 236
RSA, 236
RSCN, 236
RU, 236
run length, 236
running disparity, 236
RVSN, 236
RX_ID (responder exchange identifier), Fibre Channel, 90

S

SAN-attached storage, 236
SAN-from-legacy environment
  customer delivery of storage solution, 43
  drawbacks, 42
  Fibre Channel communication infrastructure, 41–42
  interoperability issues, 52–54
  iSCSI/Fibre Channel, 43
  NAS, 42
  ROI, 47–50
  set protocol solutions, 42
  storage architecture and design, 45–46
  storage performance surveys, 44
SANs (Storage Area Networks), 2, 236, 237
  benefits, 2–4
  business continuity, 126–127
  complexity, New York Life case study, 118
  components, 3–4
  configuration
    architecture, 40
    legacy systems, 35, 38
    system requirements analysis, 38
  differences from NAS, 31
  Fibre Channel, 6
    combined switched access and shared media, 26–27
    disaster tolerance, 28
    fabrics, switches, 8
    JBODs, 26
    loop tenancy, 28
    management facilities, 27
    management tools, 28
    NIC fabric login, 26
    RAID, 26
    shared-media SANs, 26
    SNS device discovery, 26
    switched-access networks, 26
  functional requirements, 4
  gateways, 53–54
  goals of design, 5
  HBAs, 11
  implementation, New York Life case study, 117
  interconnecting devices, 11
  iSCSI, 23
  life cycle, 39
  nationwide implementation example, 125–126
  organizations, 29
  placement in systems, 54
  project management, 55
  terminology, 29–31
  transition examples, 56–58
SAS (SAN-attached storage), 30, 237
saturated disk, 237
SBC, Project Pronto, 75
scale, 237
schema, 237
script, 237
SCSI, 237
  adapter, 237
  address, 237
  bridges, Fibre Channel, 107
  bus, 237
  parallel interface, 238
  SAN interface, 3
SCSI Enclosure Services, 238
SCSI Trade Association, 238
SDH, 238, 249
sector, 238
secure hash, 238
Secure Sockets Layer, 238
sequence initiative, 239
sequence initiator, Fibre Channel, 90, 239
sequence recipient, Fibre Channel, 90, 239
sequence status block, 239
sequences, Fibre Channel, 90, 112, 239
sequential I/O, 239
SEQ_ID (sequence identifier), Fibre Channel, 90, 238

Index

SERDES, 239
serial transmission, 239
serial adapter, 239
serial console, 239
serial SCSI, 239
serializer deserializer, 240
server, 240
Server Message Block, 240
server-based virtualization, 240
serverless backup, 240
service classes, Fibre Channel, 96–99
Service Incident Standard, 243
service level agreement, 240
SES (SCSI Enclosure Services), 28, 240
share, 240
shared-media SANs, Fibre Channel, 26
sharing drives, 37
sharing libraries. *See* library sharing.
shelf, 241
shielded enclosure, 241
SIA, 241
SID, 241
silica-based optical amplifiers, 62
simple name server, 241
Simple Network Management Protocol, 241
single (component) configuration, 241
single mode fiberoptic cable, 241
single point of failure, 241
single-ended signaling, 241
SIS, 241
SLA, 241
Small Computer System Interface, 241
small I/O request, 242
SMI, 242
SMPTE, 242
snapshot, 242
SNIA (Storage Networking Industry Association), 30, 147, 242, 246
SNIA Technology Center, 147–148
sniffer, 242
SNMP, 28, 242

SNS (Simple Name Service), Fibre Channel, 26, 242
SNW (Storage Networking World On-Line), 116
Society of Motion Picture and Television Engineers, 242
SOF (start-of-frame delimiter), Fibre Channel, 90, 243
Soft zone, 242
solicited control, 243
solicited data, 243
solid-state disk, 243
Solution Exchange Standard, 243
SONET, 70, 243
Source N_Port, 243
spare, 243
SPC (Storage Performance Council), 30
special characters, Fibre Channel, 91, 244
special code, 244
SPI, 244
spiral data transfer rate, 244
split I/O request, 244
split mirror, 244
SPOF, 244
spoofing, 244
Sprint ION (Integrated On-Demand Network), 75
SR, 244
SRM (storage resource management), 30, 244
SSA (Serial Storage Architecture), 4, 23
SSL, 248
STA, 244
stand-alone drive, 244
standard interconnect, 244
star networks, 245
start of frame, 245
static switches, Fibre Channel, 107
storage
    architecture and design, 45–46
    compression, 34

storage (*continued*)
  needs analysis, New York Life case study, 119
  performance surveys, 44
storage array, 245
storage controller, 245
storage device, 245
storage device virtualization, 245
storage domain, 246
storage element, 246
storage extent, 246
storage networking, 246
storage resource management, 246
storage subsystem virtualization, 246
storage subsystems, 30, 246
storage virtualization, 246
storage volume, 247
store and forward, 247
streamed sequence, 247
strip, 247
strip size, 247
stripe, 247
stripe depth, 247
stripe element, 247
stripe size, 247
striped array, 248
stripeset, 248
striping, 94, 248
structure of management information, 248
subdirectory, 248
subject, 248
substitution, 248
SVCs (switched virtual circuits), ATM, 74, 248
swapping, 248
switch WAN extender, Fibre Channel, 108
switch-over, 249
switchback, 249
switched fabric, Fibre Channel, 16
switched SCSI, 8
switched SSA, 8
switched virtual circuit, 249
switched-access networks, Fibre Channel SANs, 26
switches, 3, 8, 248
symmetric cryptosystem, 249
symmetric virtualization, 249
synchronization, 249
synchronous operations, 249
Synchronous Optical Network, 249
system board, 249
system disk, 250
system requirements, SAN configuration, 38
system under test, 250
S_ID, Fibre Channel, 90, 236, 243
S_Port, 236

T

T1 copy, 250
T10, 250
T11, 250
tabular mapping, 250
tape, 250
tape array, 250
tape drives, 250
  legacy backup systems, 35
  virtualization, 250
tape libraries, 2
  LAN data storage, 37
  virtualization, 250
tape transport, 250
tape virtualization, 250
Target ID, 250
targets, 3, 9, 250
TB, 251
Tbyte, 251
TCO, 251
TCP, 251
TCP/IP, 251

Index

TDS Informationstechnologie AG case study, 129
  benefits of solution, 132
    criteria for selection, 131
  high-availability storage, 130
  storage on demand, 131
tenancy, 251
tenancy of loops, Fibre Channel SANs, 28
Terabyte, 251
terminology
  Fibre Channel, 81–91
  SANs, 29–31
test system, 251
testbeds, SAN architecture development, 40
third-party copy, 251
threat, 251
throughput, 251
throughput-intensive, 251
timeserver, 251
TNC, 251
Token Ring, 252
topologies, Fibre Channel, 8, 15–16, 100, 252
total cost of ownership, 252
TPC, 252
transceiver, 252
transitioning to SANs, 56–57
  DWDM, 58
  peer models, 57
transmission characters, Fibre Channel, 91, 110, 252
transmission code, 252
Transmission Control Protocol, 252
Transmission Encode/Decode layer, FC-1, 13, 92
transmission hierarchy, Fibre Channel, 110
transmission words, Fibre Channel, 91, 111, 252
transmitter, 253
transparent fail-over, 253
trap, 253
triaxial cable, 253
Trojan horse, 253
tunneling, 253

U

UBR, ATM QoS, 74
UDP, 253
ULP, 253
ULP (upper layer protocol), Fibre Channel, 91
Ultra SCSI, 253
Ultra2 SCSI, 254
Ultra3 SCSI, 254
UML, 254
unauthorized disclosure, 254
unclassified, 254
unicast, 254
Unicode, 254
Unified Modeling Language, 255
uninterruptible power source, 255
unsolicited control, 255
unsolicited data, 255
Upper-Layer Protocol, 255
UPS, 255
usable capacity, 255
user data extent, 255
user data extent stripe depth, 255
User Datagram Protocol, 256
user identification number, 256

V

valid data byte, 256
valid frame, 256
validity control bit, 256
VBA, 256
VBR, ATM QoS, 74
VCI, 256

VCs, Fibre Channel, 99
VCSEL, 256
vendor unique, 256
verification, 256
versioning, 257
vertical cavity surface emitting laser, 257
VI (Virtual Interface), 29
VIA, 257
virtual block, 257
virtual block address, 257
virtual channel identifier, 257
virtual circuit, 257
virtual device, 257
virtual disk, 257
Virtual Interface Architecture, 258
virtual path identifier, 258
virtual tape, 258
virtualization, 258
virus, 258
volatility, 258
volume, 258
volume group, 258
volume manager, 259
volume pool, 259
volume set, 259
VPI, 259
vulnerability, 259

W

WANs, 259
   data traffic, 69
      ATM, 70–71
      QoS, 70
warm spare, 259

warm swap, 259
wave division multiplexing, 259
WBEM, 259
WDM, 65, 259
Web-based enterprise management, 259
well-known address, 260
wide area network, 260
Wide SCSI, 260
Windows driver model, 260
Windows Internet Naming Service, 260
Windows Management
      Instrumentation, 260
WINS, 260
WMI, 260
word, 260
workgroup, 261
world wide name, 261
worm, 261
write hole, 261
write penalty, 262
write-back cache, 261
write-through cache, 262
WWN, 262

X–Z

X3T10, 262
X3T11, 262
XML, 262
X_ID, 262
zero filling, 262
zeroization, 262
zone, 262
zone set, 263
zoning, 263